Fracture, Fatigue and Structural Integrity of Metallic Materials

Fracture, Fatigue and Structural Integrity of Metallic Materials

Special Issue Editors

Sergio Cicero
José Alberto Álvarez

MDPI • Basel • Beijing • Wuhan • Barcelona • Belgrade

Special Issue Editors

Sergio Cicero
Universidad de Cantabria
Spain

José Alberto Álvarez
Universidad de Cantabria
Spain

Editorial Office
MDPI
St. Alban-Anlage 66
4052 Basel, Switzerland

This is a reprint of articles from the Special Issue published online in the open access journal *Metals* (ISSN 2075-4701) from 2018 to 2019 (available at: https://www.mdpi.com/journal/metals/special_issues/fracture_structural_integrity).

For citation purposes, cite each article independently as indicated on the article page online and as indicated below:

LastName, A.A.; LastName, B.B.; LastName, C.C. Article Title. *Journal Name* **Year**, *Article Number, Page Range.*

ISBN 978-3-03928-859-5 (Pbk)
ISBN 978-3-03928-860-1 (PDF)

Contents

About the Special Issue Editors . . . . **vii**

Sergio Cicero and José Alberto Álvarez
Fracture, Fatigue, and Structural Integrity of Metallic Materials
Reprinted from: *Metals* **2019**, *9*, 913, doi:10.3390/met9080913 . . . . **1**

Cheng Yao, Zhengfei Hu, Fan Mo and Yu Wang
Fabrication and Fatigue Behavior of Aluminum Foam Sandwich Panel via Liquid Diffusion
Welding Method
Reprinted from: *Metals* **2019**, *9*, 582, doi:10.3390/met9050582 . . . . **4**

Yixin Chen, Pengmin Lv and Datao Li
Research on Fatigue Strength for Weld Structure Details of Deck with U-rib and Diaphragm in
Orthotropic Steel Bridge Deck
Reprinted from: *Metals* **2019**, *9*, 484, doi:10.3390/met9050484 . . . . **15**

David F. Hardy and David L. DuQuesnay
Effect of Repetitive Collar Replacement on the Residual Strength and Fatigue Life of Retained
Hi-Lok Fastener Pins
Reprinted from: *Metals* **2019**, *9*, 445, doi:10.3390/met9040445 . . . . **30**

Qiwei Wang, Junfeng Chen, Xiao Chen, Zengliang Gao and Yuebing Li
Fatigue Life Prediction of Steam Generator Tubes by Tube Specimens with Circular Holes
Reprinted from: *Metals* **2019**, *9*, 322, doi:10.3390/met9030322 . . . . **46**

**Martin Leitner, David Simunek, Jürgen Maierhofer, Hans-Peter Gänser and
Reinhard Pippan**
Retardation of Fatigue Crack Growth in Rotating Bending Specimens with
Semi-Elliptical Cracks
Reprinted from: *Metals* **2019**, *9*, 156, doi:10.3390/met9020156 . . . . **57**

Ryutaro Fueki, Koji Takahashi and Mitsuru Handa
Fatigue Limit Improvement and Rendering Defects Harmless by Needle Peening for High
Tensile Steel Welded Joint
Reprinted from: *Metals* **2019**, *9*, 143, doi:10.3390/met9020143 . . . . **70**

**Borja Arroyo Martínez, José AlbertoÁlvarez Laso, Federico Gutiérrez-Solana,
Alberto Cayón Martínez, Yahoska Julieth Jirón Martínez and Ana Ruht Seco Aparicio**
A Proposal for the Application of Failure Assessment Diagrams to Subcritical Hydrogen
Induced Cracking Propagation Processes
Reprinted from: *Metals* **2019**, *9*, 670, doi:10.3390/met9060670 . . . . **84**

Hye-Jin Kim, Hyeong-Kwon Park, Chang-Wook Lee, Byung-Gil Yoo and Hyun-Yeong Jung
Baking Effect on Desorption of Diffusible Hydrogen and Hydrogen Embrittlement on
Hot-Stamped Boron Martensitic Steel
Reprinted from: *Metals* **2019**, *9*, 636, doi:10.3390/met9060636 . . . . **103**

Pablo González, Sergio Cicero, Borja Arroyo and José Alberto Álvarez
Environmentally Assisted Cracking Behavior of S420 and X80 Steels Containing U-notches at
Two Different Cathodic Polarization Levels: An Approach from the Theory of Critical Distances
Reprinted from: *Metals* **2019**, *9*, 570, doi:10.3390/met9050570 . . . . **117**

Ali Reza Torabi, Filippo Berto and Alberto Sapora
Finite Fracture Mechanics Assessment in Moderate and Large Scale Yielding Regimes
Reprinted from: *Metals* **2019**, *9*, 602, doi:10.3390/met9050602 . . . . **132**

Sergio Cicero, Juan Diego Fuentes, Isabela Procopio, Virginia Madrazo and Pablo González
Critical Distance Default Values for Structural Steels and a Simple Formulation to Estimate the
Apparent Fracture Toughness in U-Notched Conditions
Reprinted from: *Metals* **2018**, *8*, 871, doi:10.3390/met8110871 . . . . **142**

Yoshinobu Shimamura, Shinya Matsushita, Tomoyuki Fujii, Keiichiro Tohgo, Koichi Akita,
Takahisa Shobu and Ayumi Shiro
Feasibility Study on Application of Synchrotron Radiation μCT Imaging to Alloy Steel for Non-
Destructive Inspection of Inclusions
Reprinted from: *Metals* **2019**, *9*, 527, doi:10.3390/met9050527 . . . . **152**

About the Special Issue Editors

Sergio Cicero holds an MSc in Civil Engineering (2002), a PhD in Civil Engineering (2007), and a MSc in Business Management (2007). His main areas of expertise are structural integrity and mechanical characterization, covering the four main failure modes (fracture, fatigue, creep, and corrosion). He has conducted research stays at Swinden Technology Centre (Rotherham, UK) and GKSS Forschungszentrum (Geestacht, Germany), participated in several European projects (e.g., FITNET, HIPERCUT, INCEFA PLUS), and has published over 75 papers in international journals with JCR impact factors. He is currently Full Professor of Materials Science and Metallurgical Engineering at the University of Cantabria, where he is also Director of Research.

José A. Álvarez is Full Professor of Materials Science and Metallurgical Engineering at the University of Cantabria. He obtained his PhD in Mechanical Engineering in 1996 and his main areas of expertise are in the analysis of damage and environmentally assisted cracking processes as well as the characterization of metallic materials. He has broad experience in European projects and has published over one hundred scientific papers, many of them in international indexed journals.

Editorial

Fracture, Fatigue, and Structural Integrity of Metallic Materials

Sergio Cicero * and José Alberto Álvarez

LADICIM (Laboratory of Materials Science and Engineering), University of Cantabria, E.T.S. de Ingenieros de Caminos, Canales y Puertos, Av/Los Castros 44, 39005 Santander, Spain
* Correspondence: ciceros@unican.es; Tel.: +34-942200917

Received: 15 August 2019; Accepted: 20 August 2019; Published: 20 August 2019

---

1. Introduction and Scope

Fracture, fatigue, and other subcritical processes, such as creep crack growth or stress corrosion cracking, present numerous open issues from both scientific and industrial points of view. These phenomena are of special interest in industrial and civil metallic structures, such as pipes, vessels, machinery, aircrafts, ship hulls, and bridges, given that their failure may imply catastrophic consequences for human life, the natural environment and/or the economy. Moreover, an adequate management of their operational life, defining suitable inspection periods, repairs, or replacements, requires their safety or unsafety conditions to be defined.

The analysis of these technological challenges requires accurate comprehensive assessments tools based on solid theoretical foundations, as well as structural integrity assessment standards or procedures incorporating such tools into industrial practice.

This Special Issue is focused on new advances in fracture, fatigue, creep and corrosion analysis of metallic structural components containing defects (e.g., cracks, notches, metal loss, etc.), and also on those developments that are being or could be incorporated to structural integrity assessment procedures.

2. Contributions

Twelve research contributions (eleven articles and one communication) have been published in this Special Issue. Eleven contributions deal with critical or subcritical phenomena, and one is related to non-destructive inspections. Among the former, six of them provide significant advances on fatigue research, three of them deal with the analysis of hydrogen issues and their effect on the mechanical behavior, and two are related to the fracture analysis of notches.

Thus, in the fatigue context, Yao et al. [1] analyze the fatigue behavior of aluminum foam sandwich panels fabricated through liquid diffusion welding and adhesive methods, also providing microstructural and metallurgical analyses; Chen et al. [2] investigate the fatigue behavior of a given steel bridge deck by characterizing (experimentally) the fatigue performance of the deck plates and the deck welded details, and performing stress analysis. Hardy and DuQuesnay [3] characterize the fatigue performance of hi-lok fasteners used in aircraft structural joints and subjected to multiple collar replacements, as well as the corresponding behavior under static loading conditions. Wang et al. [4] provide predictions of the fatigue life of Inconel 690 steam generator tubes by using tubular specimens containing holes. Leitner et al. [5] analyze how overloads induce fatigue crack growth retardation on EA47 steel round bars containing semi-elliptical cracks, comparing the experimental results to the predictions provided by a modified NASGRO equation. Finally, Fueki et al. [6] evaluates the fatigue limit improvement caused by needle peening in high tensile (strength) steel HT780. All these contributions cover different sectors, such as aerospace, railway, bridge design, or energy generation, among others.

The three contributions related to hydrogen research start with Arroyo et al. [7], who provide a new methodology to analyze subcritical Hydrogen Induced Cracking by using Failure Assessment Diagrams, a tool that is commonly used in fracture-plastic collapse structural integrity assessments. Then, Kim et al. [8] study the role of diffusible hydrogen on delayed fractures in hot-stamped ultrahigh strength steels used in automotive structural components, and analyze how baking times and temperatures affect the cracking behavior. Lastly, González et al. [9] analyze the Environmentally Assisted Cracking (Hydrogen Embrittlement) behavior of two steels containing notch-type defects. Their analysis is based on the Theory of Critical Distances, a theory that had only been used to analyze fracture and fatigue phenomena.

Finally, the two contributions related to the fracture analysis of notch-type defects are those provided by Torabi et al. [10] and Cicero et al. [11], respectively. The former investigates, by using Finite Fracture Mechanics and the Equivalent Material Concept, the ductile fracture initiation of two aluminum alloys containing different notch types. The latter provides default values of the critical distance for structural steels operating at their corresponding Lower Shelf and Ductile-to-Brittle Transition Zone, and provides formulation for the apparent fracture toughness estimation of such steels when containing U-shaped notches.

The final contribution [12] is a communication that examines the feasibility of applying synchrotron radiation μCT imaging for non-destructive inspection of inclusions in alloy steels.

3. Conclusions and Outlook

The contributions of this Special Issue provide different advances in fracture, fatigue, and structural integrity research. Their application affects a number of engineering sectors, such as aerospace, mechanical, civil, railway, materials, and energy, among others. Evidently, there are still numerous open issues to solve in this context, and engineering applications still need to be more developed and improved, but as guest editors, we hope this Special Issue provides a significant impact and also that the scientific and engineering communities found it interesting.

Finally, we would like to thank all the authors for their contributions, and all the reviewers for their outstanding efforts to improve the scientific quality of the different documents composing this this Special Issue. We would also like to give special thanks to all the staff at the Metals Editorial Office, especially to Betty Jin, who managed and simplified the publication process.

Conflicts of Interest: The authors declare no conflicts of interest.

References

1. Yao, C.; Hu, Z.; Mo, F.; Wang, Y. Fabrication and Fatigue Behavior of Aluminum Foam Sandwich Panel via Liquid Diffusion Welding Method. *Metals* **2019**, *9*, 582. [CrossRef]
2. Chen, Y.; Lv, P.; Li, D. Research on Fatigue Strength for Weld Structure Details of Deck with U-rib and Diaphragm in Orthotropic Steel Bridge Deck. *Metals* **2019**, *9*, 484. [CrossRef]
3. Hardy, D.F.; DuQuesnay, D.L. Effect of Repetitive Collar Replacement on the Residual Strength and Fatigue Life of Retained Hi-Lok Fastener Pins. *Metals* **2019**, *9*, 445. [CrossRef]
4. Wang, Q.; Chen, J.; Chen, X.; Gao, Z.; Li, Y. Fatigue Life Prediction of Steam Generator Tubes by Tube Specimens with Circular Holes. *Metals* **2019**, *9*, 322. [CrossRef]
5. Leitner, M.; Simunek, D.; Maierhofer, J.; Gänser, H.P.; Pippan, R. Retardation of Fatigue Crack Growth in Rotating Bending Specimens with Semi-Elliptical Cracks. *Metals* **2019**, *9*, 156. [CrossRef]
6. Fueki, R.; Takahashi, K.; Handa, M. Fatigue Limit Improvement and Rendering Defects Harmless by Needle Peening for High Tensile Steel Welded Joint. *Metals* **2019**, *9*, 143. [CrossRef]
7. Arroyo Martínez, B.; Álvarez Laso, J.A.; Gutiérrez-Solana, F.; Cayón Martínez, A.; Jirón Martínez, Y.J.; Seco Aparicio, A.R. A Proposal for the Application of Failure Assessment Diagrams to Subcritical Hydrogen Induced Cracking Propagation Processes. *Metals* **2019**, *9*, 670. [CrossRef]
8. Kim, H.J.; Park, H.K.; Lee, C.W.; Yoo, B.G.; Jung, H.Y. Baking Effect on Desorption of Diffusible Hydrogen and Hydrogen Embrittlement on Hot-Stamped Boron Martensitic Steel. *Metals* **2019**, *9*, 636. [CrossRef]

9. González, P.; Cicero, S.; Arroyo, B.; Álvarez, J.A. Environmentally Assisted Cracking Behavior of S420 and X80 Steels Containing U-notches at Two Different Cathodic Polarization Levels: An Approach from the Theory of Critical Distances. *Metals* **2019**, *9*, 570. [CrossRef]
10. Torabi, A.R.; Berto, F.; Sapora, A. Finite Fracture Mechanics Assessment in Moderate and Large Scale Yielding Regimes. *Metals* **2019**, *9*, 602. [CrossRef]
11. Cicero, S.; Fuentes, J.D.; Procopio, I.; Madrazo, V.; González, P. Critical Distance Default Values for Structural Steels and a Simple Formulation to Estimate the Apparent Fracture Toughness in U-Notched Conditions. *Metals* **2018**, *8*, 871. [CrossRef]
12. Shimamura, Y.; Matsushita, S.; Fujii, T.; Tohgo, K.; Akita, K.; Shobu, T.; Shiro, A. Feasibility Study on Application of Synchrotron Radiation μCT Imaging to Alloy Steel for Non-Destructive Inspection of Inclusions. *Metals* **2019**, *9*, 527. [CrossRef]

Article

Fabrication and Fatigue Behavior of Aluminum Foam Sandwich Panel via Liquid Diffusion Welding Method

Cheng Yao, Zhengfei Hu *, Fan Mo and Yu Wang

Shanghai Key Lab for R&D and Application of Metallic Functional Materials, School of Materials Science and Engineering, Tongji University, Shanghai 201804, China; yaocheng1230@tongji.edu.cn (C.Y.); mofan61@163.com (F.M.); 17717099406@163.com (Y.W.)
* Correspondence: huzhengf@tongji.edu.cn; Tel.: +86-138-1801-9882

Received: 30 April 2019; Accepted: 10 May 2019; Published: 20 May 2019

Abstract: Aluminum Foam Sandwich panels were fabricated via liquid diffusion welding and glue adhesive methods. The Microstructure of the Aluminum Foam Sandwich joints were analyzed by Optical Microscopy, Scanning Electron Microscopy, and Energy Dispersive Spectroscopy. The metallurgical joints of Aluminum Foam Sandwich panels are compact, uniform and the chemical compositions in the diffusion transitional zone are continuous, so well metallurgy bonding between Aluminum face sheet and foam core was obtained. The joining strength of an Aluminum Foam Sandwich was evaluated by standard peel strength test and the metallurgical joint Aluminum Foam Sandwich panels had a higher peel strength. Moreover, a three-point bending fatigue test was conducted to study the flexural fatigue behavior of Aluminum Foam Sandwich panels. The metallurgical joint panels have a higher fatigue limit than the adhesive joining sandwich. Their fatigue fracture mode are completely different, the failure mode of the metallurgical joint is faced fatigue; the failure mode for the adhesive joint is debonding. Therefore, the higher joining strength leads to a longer fatigue life.

Keywords: aluminum foam sandwich; microstructure; three-point bending fatigue; peel strength

---

1. Introduction

Metallic foams, especially Aluminum foams (AF), have recently received attention because of their outstanding physical and chemical properties, including low density, high specific strength, impacting energy absorption, sound absorption, flame resistance and electromagnetic shield effectiveness [1–8]. Due to these aforementioned capabilities, metal foams can be used for many industrial applications such as aerospace, marine, railway, automotive and construction [9–11].

Aluminum Foam Sandwich (AFS) is a special class of composites materials which is widely used for panels, shells, tubes, crash protection devices and lightweight structures [12,13]. It is fabricated by sandwiching a thick AF as core material between two thin alloy sheets as facing sheets. With this sandwich structure, the AFS can provide specific strength, better dimensional stability, improved damping and acoustic insulation properties compared with the simply AF [14,15], as the core foam bears shear load meanwhile the face sheets carry an axial load and resist against bending [16].

Different joining techniques have been developed to join the AF core and facing sheets. Though adhesive is the most common method with low cost and simple operation [17], adhesive AFS has some drawbacks such as low joining strength and low-temperature resistance. In order to improve the joining strength and temperature resistance, metallurgical joining techniques such as casting, brazing [18] and soldering [19] are developed. Besides these traditional techniques, some other joining techniques based on metallurgical joining also have been proposed and investigated feasibility.

Tungsten inert gas welding is a welding method for generating electric arc from activating tungsten or pure tungsten with inert gas to protect the manufacturing environment. This method has the benefits of good operability and low cost [20]. The Fluxless soldering technique, with a filling material SnAg4Ti working at 220–229 °C, can make the solder spread under the Oxide layer of the metal and completely wet the surface of the metal substrate. Then, the Oxide layer around the molten solder is removed mechanically, and the atoms of the solder and the substrate are closely joint [21,22]. Besides, metal glass brazing is also promoted to be a potential method with the soldering material of amorphous alloy [23]. Moreover, the sandwich and foam filled tubes made of Aluminum alloys could be fabricated by powder metallurgy method. The joining between the metal foams and the tubes or sheet are achieved during the formation of the liquid metallic foam, promoting a metallic bonding without any joining step. Results have demonstrated that the thermal treatment that is submitted to these tubes or sheets at high temperatures during the foam formation is beneficial to obtain a predictable and stable mechanical behavior of the resulting in-situ foam structures [24]. The ductility of these structures increases, leading to an efficient crashworthiness without the formation of cracks and abrupt failure when subjected to compressive and bending loads.

Among the metallurgical joining techniques, diffusion welding technique is commonly applied to prepare metals and complex structures. AFS panel prepared via this method has good mechanical properties because the metal microstructures of the core substrate and the face panels on the interface are compact and continuous. Kitazono et al. [25] studied the diffusion welding technique for closed-cell Al foam, and suggested that in the process of joint formation, the compressive stress is strong enough to break the Oxide film on the contacting surface, which reduces defects and promotes the diffusion of Al atom. Bangash et al. [26] have investigated the joining area of AF core to Al alloy sheets with Zn-based joining material and found that in the joining area, the presence of Al rich and Zn rich phases confirm the diffusion, ensuring strong metallurgical joining. They indicated that the joining process can easily be automated in a continuous furnace, guaranteeing high productivity, reproducibility and cheap industrial cost. However, they do not focus on the manufacture of large size AFS panels, which is one of the purposes of this paper.

In this paper, AFS beams were prepared by a specially designed method via a liquid diffusion welding process, which achieved a high strength joining between the AF core and face sheets. The fatigue behavior under three-point bending cyclic load is investigated. The effects of joining strength to fatigue behavior is analyzed, and fatigue fracture characteristics are discussed as well.

2. Materials and Experimental Methods

Aluminum alloy (AA)-5056 (Al 94.8% + Mg 4.5% + Fe 0.4% + Si 0.3%) plate (density 2.7 g/cm^3), 1.0 mm thick, was selected to be the face sheet material for AFS. A lightweight, non-flammable eco-friendly closed-cell pure AF plate (average cell size 7 mm, average density 0.4 g/cm^3, porosity 85%), 25 cm thick, produced by Yuantaida, Sichuan, China, was used as the core material. For the soldering material, Zn + 10% Al (Zn-10Al), with the liquid-solid phase transition temperature of 426.5 °C and 380.0 °C, was chosen to be the joint alloy because of its proper molten range of the Zinc-Aluminum binary alloy, good mechanical properties and wetting properties to the Aluminum alloy substrate. It was made from pressure cast 3# Zinc alloy (Zn 95.7% + Al 4.3%) and industrial pure Aluminum (Al ≥99.5%) by ourselves.

The Al alloy sheet and AF plate were all cut off in pieces of length 300 mm, width 50 mm, thickness 1 mm (sheet) and 25 mm (AF). The length-thickness ratio of the foam core referred to ASTM C393 [27]. The joining surfaces of the AF core and sheets were abraded with 200 mesh SiC paper in order to clean the Al Oxide and facilitate the metal joining process. Then the surfaces were cleaned with alcohol in an ultrasonic bath for 5 min.

The joining surfaces of AF core and face sheets were put into the melting solder bath about 450 °C for hot-dip coating for 50 s to form the joining alloy on the joining surfaces. Finally, the pre-coated AF core and sheets were jointed by a heat press process assisted with ultrasonic vibration 1 min to establish the optimal soldering parameters and achieve high joining strength. During the heat press process, induction heating equipment was employed to heat the sample up to 430 °C to remelt the solder coating.

For comparison purposes, the adhesive joining AFS panels were also fabricated by a heat compress method at 80 °C for 5 min with aerial adhesive film as joining glue. The macromorphology of two kinds of AFS panels from different manufacturing processes are shown in Figure 1.

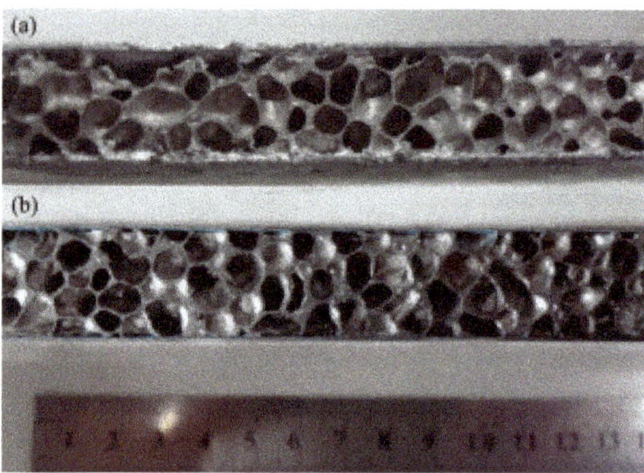

Figure 1. Two kinds of AFS (Aluminum Foam Sandwich) panels from different manufacturing process (**a**) metallurgical joining (**b**) adhesive joining.

The AFS samples made for the mechanical tests were 300 mm long, 50 mm wide and 27 mm in thickness. Peel strength test was carried out by WDW-10 universal testing machine (Loading speed 25 mm/min) to test the joining strength of the two kinds of AFS samples according to the GB T1457-2005 [28]. Three-point bending fatigue test was carried out by MTS-809 fatigue test machine with the self-made fixture (Span length 200 mm), as shown in Figure 2, according to the ASTM C393 [27]. The AFS samples were tested at loading ratios $R = -0.1$, cycle frequency $C = 7$ Hz. A digital camera was used to record the fracture process to describe the fatigue behavior.

Figure 2. Loading diagram of sample and fixture for fatigue test.

3. Results and Discussion

3.1. Microstructure

Figure 3 shows the two kinds of AFS joining interface microstructure. As shown in Figure 3a,b, the seam of metallurgical joint is a soldering Zn-Al alloy which have a typical eutectic structure. α-Al dendrites nucleated on the substrate surfaces of the Al sheet and foam core and grew into the Zn-Al alloy fusion area. This indicates that the molten ZnAl alloy have good wettability on the substrate surface and the interfaces of the ZnAl joining alloy and the substrates are intimate. Furthermore, the fusion is compact and continuous without any obvious macroscopic defects. Figure 3c,d present that the joint interface of the adhesive AFS sample is a typical physical joint with many obvious defects such as holes and irregular inclusions, which might be formed during the adhering process. The air between Al alloy sheets and Al foam could not be all pulled out, and the remain air in the adhesive might form the air holes when adhesive melts. Moreover, during the melting and remodeling process, the thermoplastic glue film would shrink and also form hole defects. The final result is that the physical joint is irregular and the connection between adhesive and substrates might be not tight, which lead to poor strength of the adhesive joint.

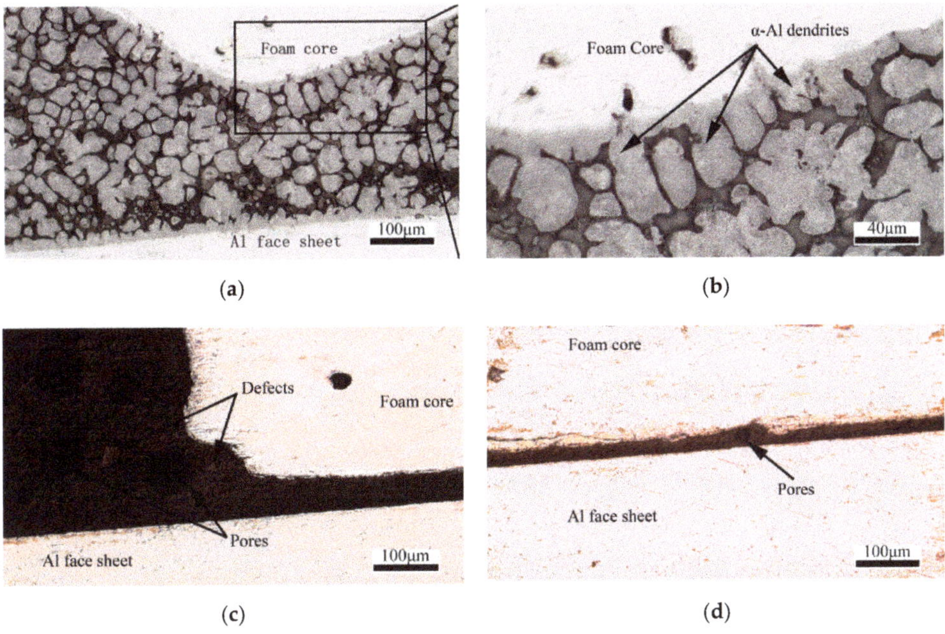

Figure 3. Optical microscope images of AFS joining parts (**a,b**) Metallurgical joining (**c,d**) Adhesive joining.

The elemental composition of the interface between the Zn-Al alloy fusion and the substrates of Al sheet and foam core is shown in Figure 4. The main elements across the interface (Al in sheets and foam core, Zn in Zn-Al alloy) are continuous. It indicates that the mutual diffusion of Aluminum and Zinc atoms occurred in the process of metallurgical joining. Figure 4c,d shows that the contents of Zinc atoms and Aluminum atoms change linearly and continuously from the Zinc-rich seam to the Aluminum-rich area within a certain diffusion distance. The molten Zn-10Al alloy has good wettability and diffusion on the surface of the Aluminum substrate during hot-dip coating and metallurgical joining process. The inter-diffusion greatly improves the joining strength.

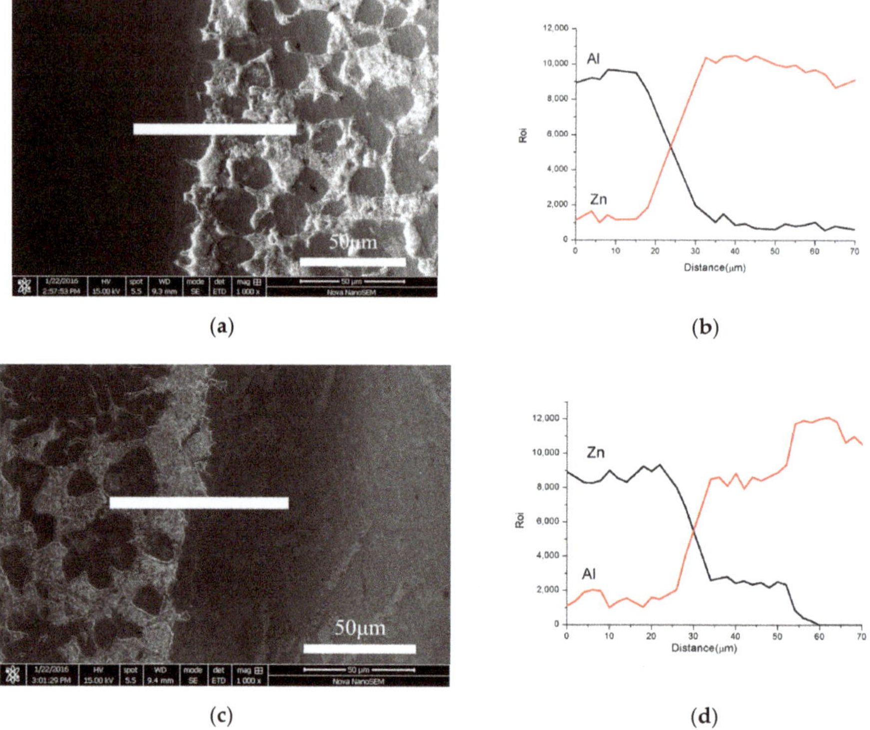

Figure 4. Scanning Electron Microscopy (SEM) images of a metallurgical joint (**a**) Interface of Al face sheet and Zn-10Al alloy (**b**) Energy Dispersive Spectroscopy (EDS) line analysis of Figure 4a (**c**) Interface of Zn-10Al alloy and Aluminum foam (**d**) EDS line analysis of Figure 4c.

The surfaces of Al sheet and AF core are pretreated to remove dirt and Aluminum Oxide before fabrication. In the process of hot-dipping, the joint interfaces of AF and Al sheets are in direct contact with the molten Zn-10Al alloy. Al and Zn have high solid solubility to each other at high temperature, which result to obvious mutual diffusion behavior [29,30]. Zn atom diffuse from the alloy to the sheets and core substrates, while Al atom diffuse to the opposite side. The higher the Zn content in the Al interface, the lower the melt point of Al alloy in joint interface. The melting parts of the surface cause the oxide film broken and destroyed. In conclusion, the mutual diffusion of alloy atom and the remove of oxide film promote the combination of alloy coating and lays a good foundation for AFS manufacture.

During the hot press process, the Zinc-based alloy remelt at a high temperature. Since the oxide film on the surface of the substrates is partially removed in the pretreatment, the Zinc atoms have good wettability to the substrates [29]. They easily diffuse into the Aluminum substrate and occupy the position of the oxide film and some Aluminum atoms, which lead to more frequent mutual diffusion. Similar to the hot-dip process, the higher the Zn content in the Al interface, the lower the melt point of Al alloy in the joint interface. Eventually, part of the Aluminum substrate under the oxide film melt. Since the oxide film had a lower density than the alloy, it floated in the molten alloy and separated from the Aluminum substrate. Ultrasound makes the Al foam and Al sheets vibrate to remove the floating oxide film and promote the diffusion effect in the molten area. Former studies [30,31] showed that when ultrasound is applied to a metal melt, it will caused the cavitation effect. The cavitation effect induce mechanical effects such as acoustic flow and shock waves in the metal melt [32,33]. These mechanical effects also can destroy the oxide film and help to achieve completely wetting. Moreover, the mechanical effects caused by the cavitation effect will also play a role similar to stirring [34], eliminating the inclusion of air in the molten Zi-Al alloy. Finally, the fusion area is compact and continuous without any visible defects.

3.2. Peel Strength Test

In order to verify the research of microstructure, peel strength test was carried out to check the joining effects. Three samples from each the two kinds of AFS panels were tested. The results reported in Figure 5 show that the average peel strength of metallurgical joining samples (140.0 N·mm/mm) is higher than the adhesive joining samples (27.5 N·mm/mm) (see Table 1).

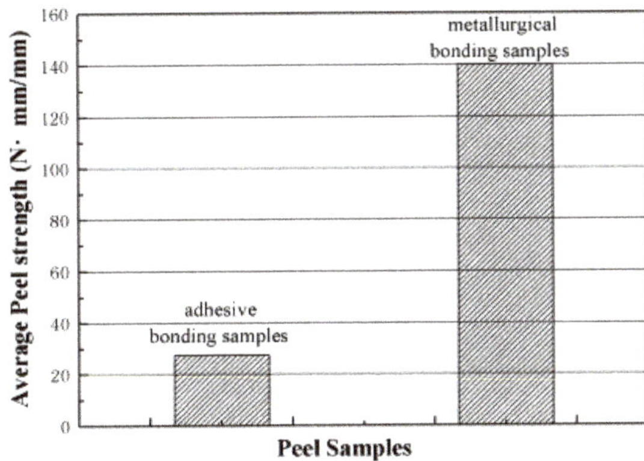

Figure 5. Average peel strength of two types of samples.

Table 1. Experimental data for peel strength.

Samples	Joint	Peel Strength (N·mm/mm)	Average
1	adhesive	25.2	
2	adhesive	28.3	27.5
3	adhesive	29.0	
4	metallurgical	147.3	
5	metallurgical	135.1	140.0
6	metallurgical	137.6	

Figure 6 shows the fracture morphology of two types of AFS samples after the peel strength test. For the metallurgical joining AFS, the main destroyed part was not the hot-dip coating but the Al foam core, as shown in Figure 6b. That means the strength of the joint is higher than the Al-foam core. In contrast, for the adhesive ones, the glue film and the sheets were nearly completely detached, which means the strength of the joint is lower than the film itself and the AF core. It may result from the defects made during the heat press process in the adhesive area.

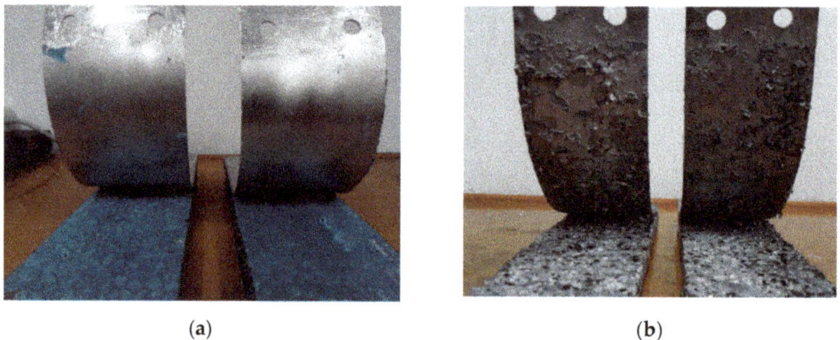

(a)   (b)

Figure 6. Appearances of Peel Fractures of two types of samples (**a**) Adhesive joining samples (**b**) Metallurgical joining samples.

3.3. Three-Point Bending Fatigue

Figure 7 shows the S-N curves of the two kinds of sandwich panels. Under the same load, the fatigue life of metallurgical joining samples is much longer than the adhesive ones. Harte et al. [35] investigated the AFS four-point fatigue behavior and fitted the S-N curve to predict the fatigue limit. With this method, the S-N curves are fitted with the average fatigue life (three repeated tests for every given load, in Table 2) of every given load to achieve the experience formula of metallurgical joining samples (Equation (1)) and adhesive joining samples (Equation (2)).

$$S = 7380 - 645.2 \log N \tag{1}$$

$$S = 6162 - 497.5 \log N. \tag{2}$$

Since N is cycles, we define the fatigue limit as the force when fatigue life is about 5 million cycles [36]. According to Equations (1) and (2) above, the fatigue limit of metallurgical joining samples is 3058 N, adhesive joining samples is 2829 N (see Table 2). Besides, since it is the estimating limits, for further research more factors should be considered such as the strain degradation of pure foams [37].

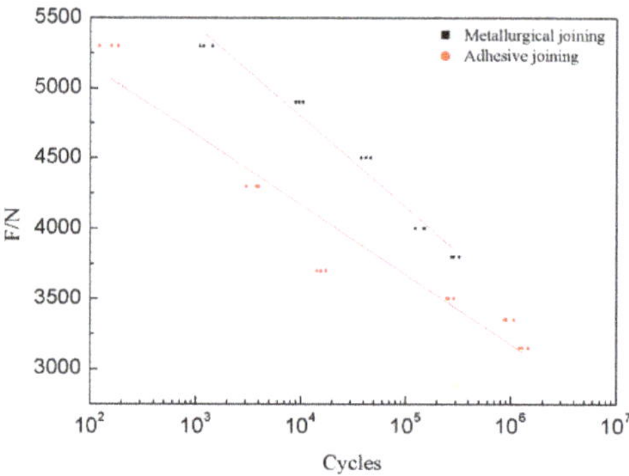

Figure 7. S-N curves of metallurgical joining and adhesive joining AFS.

The deflection curves of the two kinds of samples for various stress levels are shown in Figure 8. At the beginning, the fatigue response comprised a slow rate of accumulation of deflection with increasing cycles. The defects in the incubation period grew slowly and steadily due to the microdamage development. When microdamage developed to a critical collapsing level, the inner crake rapidly grew and the rate of deflection increased dramatically, which means the end of the incubation period and the coming of failure. Comparing Figure 8a with Figure 8b, the critical collapsing level of the metallurgical joining samples is much higher than the adhesive joining samples at the same loading force. These behaviors above are similar to that noted by Harte et al. [35].

Table 2. Experimental data for three-point bending fatigue.

Samples	Joint	Max Loading (N)	Life Cycles	Average
1	metallurgical	5300	1438	
2	metallurgical	5300	1092	1238
3	metallurgical	5300	1184	
4	metallurgical	4900	8890	
5	metallurgical	4900	10,435	9639
6	metallurgical	4900	9592	
7	metallurgical	4500	37,623	
8	metallurgical	4500	46,071	41,881
9	metallurgical	4500	41,949	
10	metallurgical	4000	147,892	
11	metallurgical	4000	152,377	141,186
12	metallurgical	4000	123,289	
13	metallurgical	3800	325,071	
14	metallurgical	3800	289,765	296,950
15	metallurgical	3800	276,014	
16	adhesive	5300	123	
17	adhesive	5300	186	157
18	adhesive	5300	162	
19	adhesive	4300	3981	
20	adhesive	4300	3032	3592
21	adhesive	4300	3763	
22	adhesive	3700	15,769	
23	adhesive	3700	14,291	15,861
24	adhesive	3700	17,523	
25	adhesive	3500	260,817	
26	adhesive	3500	291,082	267,345
27	adhesive	3500	250,136	
28	adhesive	3350	883,188	
29	adhesive	3350	1,078,321	958,252
30	adhesive	3350	913,247	
31	adhesive	3150	1,213,484	
32	adhesive	3150	1,456,872	1,320,030
33	adhesive	3150	1,289,734	

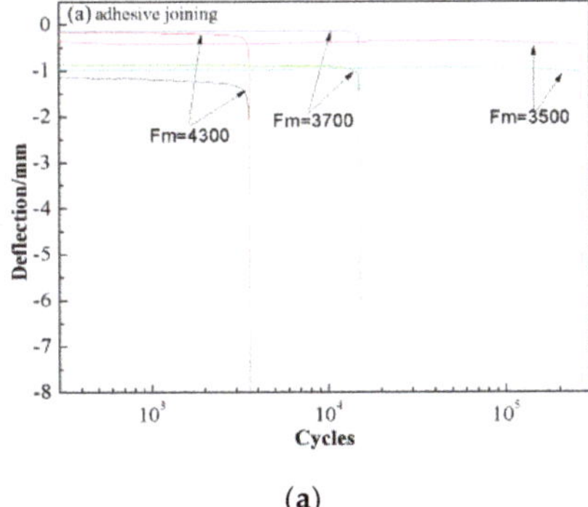

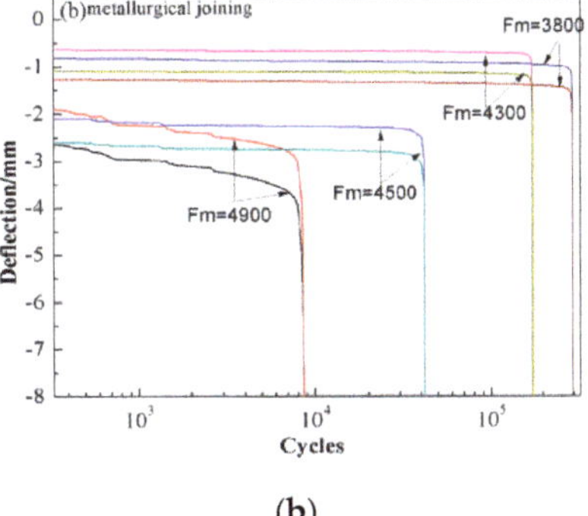

(a)   (b)

Figure 8. Deflection versus number of cycles of two kinds of samples (**a**) adhesive joining samples (**b**) metallurgical joining samples.

In former studies, Harte et al. [35,38] suggested three types of bending fatigue failure modes for sandwich structure: face fatigue, indentation fatigue and core shear. Face fatigue is manifested the fatigue cracks across the face sheet on the tensile side of the beam. Indentation fatigue appears to be the indentation beneath the loading rollers. In addition, core shear displays the fatigue cracks inclined at an angle of approximately 45° to the neutral axis. In addition, if the face sheets change to higher strength material, such as AISI 304 steel, the AFS panels will fail in only two modes, indentation fatigue and core shear [39].

The fatigue fractures of the two kinds of AFS panels are shown in Figure 9. Since the fatigue cracks are on the tensile side of AFS panels, it is clear that the failure mode of adhesive AFS is debonding and face fatigue, while the failure mode of a metallurgical AFS is face fatigue without debonding. According to the test records, the failure process of adhesive samples can be divided into two stages. First the core and the face sheets debonded, then the Aluminum foam core was broken by the bending force. When debonding occurs, the foam core will loss the protect of face sheet and quickly achieve the collapse load.

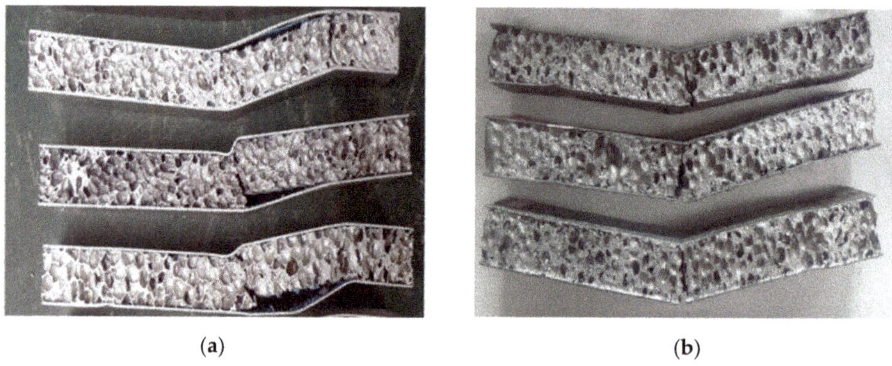

(a)   (b)

Figure 9. Fatigue fracture morphology of two types of AFS (**a**) adhesive joining samples (**b**) metallurgical joining samples.

For face fatigue, the peak load F is

$$F = \frac{4bt(c+t)}{l-S}\sigma_f + \frac{bc^2}{1-S}\sigma_c. \tag{3}$$

For Equation (3) [37], b is the AFS width, t is the thickness of face sheet, l is the span length of three-point bending, S is the length between to upper forces in four-point bending fatigue failure, σ_f is the yield strength of face sheet, σ_c is the yield strength of core.

While in three-point bending, S can be regarded as zero, so Equation (3) can change to:

$$F = \frac{4bt(c+t)}{l}\sigma_f + \frac{bc^2}{l}\sigma_c. \tag{4}$$

When debonding occurs, the face sheets and core are separated, and the peak load will be the smaller one of the two parts in the formula. So we can say that the face sheet debonding lead to a decrease of bending strength. In conclusion, the metallurgical joining did not show the process of debonding, which shows obviously the higher bonding strength.

4. Conclusions

The metallurgical joining of AFS was successfully fabricated via the method of hot-dipping, hot-press process with ultrasonic vibration assisted, Zn-10Al alloy used as a soldering material.

The ultrasonic vibration makes the molten solder compact and continuous, and removes the oxide film, which leads to the high strength joint. The average peel strength of metallurgical joining is 140.0 N·mm/mm, which is much higher than the adhesive joining. The metallurgical joining AFS have a much longer fatigue life than the adhesive sample under the same load, and also the fatigue limit of metallurgical joining AFS is much higher than adhesive joining AFS. The fatigue fracture mode for these two types of structure is completely different. The failure mode of the metallurgical joining is face fatigue, while the failure mode for the adhesive sample is face fatigue and debonding. The higher joining strength leads to a fatigue life improved significantly.

Author Contributions: Conceptualization: Z.H.; formal analysis: C.Y.; funding acquisition: Z.H. and F.M.; investigation: C.Y. and Y.W.; methodology: C.Y.; supervision: Z.H. and F.M.; validation, C.Y. and Y.W.; writing—original draft: C.Y.; writing—review and editing: Z.H.

Funding: This research was funded by the National Key Research and Development Program of China (Project No. 2017YFB0103700, 2013BAG19B01).

Conflicts of Interest: The authors declare no conflict of interest.

References

1. Ashby, M.F.; Lu, T.J. Metals foam: A survey. *Sci. China* **2003**, *46*, 521–532. [CrossRef]
2. Davies, G.J.; Zhen, S. Metallic foams: their production, properties and applications. *J. Mater. Sci.* **1983**, *18*, 1899–1911. [CrossRef]
3. Alvandi-Tabrizi, Y.; Whisler, D.A.; Kim, H.; Rabiei, A. High strain rate behavior of composite metal foams. *Mater. Sci. Eng.* **2015**, *A631*, 248–257. [CrossRef]
4. Gui, M.C.; Wang, D.B.; Wu, J.J.; Yuan, G.J.; Li, C.G. Deformation and damping behaviors of foamed Al-Si-SiCp composite. *Mater. Sci. Eng.* **2000**, *A286*, 282–288. [CrossRef]
5. Gibson, L.J.; Ashby, M.F. *Cellular Solids: Structure & Properties*; Cambridge University Press: Cambridge, UK, 1997; p. 510.
6. Katona, B.; Szebényi, G.; Orbulov, I.N. Fatigue properties of ceramic hollow sphere filled aluminium matrix syntactic foams. *Mater. Sci. Eng.* **2017**, *A679*, 350–357.
7. Katona, B.; Szebényi, A.T.; Orbulov, I.N. Compressive characteristics and low frequency damping of aluminium matrix syntactic foams. *Mater. Sci. Eng.* **2019**, *A739*, 140–148.
8. Vendra, L.; Nevile, B.; Rabiei, A. Fatigue in aluminium-steel and steel-steel composite foams. *Mater. Sci. Eng.* **2009**, *A517*, 146–153.
9. Banhart, J.; Seeliger, H.W. Aluminium Foam Sandwich Panels: Manufacture, Metallurgy and Applications. *Adv. Eng. Mater.* **2008**, *10*, 793–802. [CrossRef]
10. Garcia-Moreno, F. Commercial Applications of Metal Foams: Their Properties and Production. *Materials* **2018**, *9*, 85. [CrossRef]
11. Crupi, V.; Epasto, G.; Guglielmino, E. Comparison of aluminium sandwiches for lightweight ship structures: Honeycomb vs. foam. *Mar. Struct.* **2013**, *30*, 74–96. [CrossRef]
12. Golovin, I.S.; Sinning, H.R. Damping in some cellular metallic materials. *J. Alloys Compd.* **2003**, *355*, 2–9. [CrossRef]
13. Yang, D.H.; Yang, S.R.; Wang, H.; Ma, A.B.; Jiang, J.H.; Chen, J.Q.; Wang, D.L. Compressive properties of cellular Mg foams fabricated by melt-foaming method. *Mater. Sci. Eng.* **2010**, *A527*, 5405–5409. [CrossRef]
14. Huang, Z.; Qin, Z.; Chu, F. Damping mechanism of elastic-viscoelastic-elastic sandwich structures. *Compos. Struct.* **2016**, *153*, 96–107. [CrossRef]
15. Yu, G.C.; Feng, L.J.; Wu, L.Z. Thermal and mechanical properties of a multifunctional composite square honeycomb sandwich structure. *Mater. Des.* **2016**, *102*, 238–246. [CrossRef]
16. Harte, A.M.; Fleck, N.A.; Ashby, M.F. Sandwich panel design using Aluminum alloy foam. *Adv. Eng. Mater.* **2000**, *2*, 219–222. [CrossRef]
17. Hangai, Y.; Takahashi, K.; Yamaguchi, R.; Utsunomiya, T.; Kitahara, S.; Kuwazuru, O.; Yoshikawa, N. Nondestructive observation of pore structure deformation behavior of functionally graded Aluminum foam by X-ray computed tomography. *Mater. Sci. Eng.* **2012**, *A556*, 678–684. [CrossRef]

18. Chen, N.; Feng, Y.; Chen, J.; Li, B.; Chen, F.; Zhao, J. Vacuum Brazing Processes of Aluminum Foam. *Rare Metal Mater. Eng.* **2013**, *42*, 1118–1122.

19. Matsumoto, R.; Tsuruoka, H.; Otsu, M.; Utsunomiya, H. Fabrication of skin layer on Aluminum foam surface by friction stir incremental forming and its mechanical properties. *J. Mater. Process. Technol.* **2015**, *218*, 23–31. [CrossRef]

20. D'Urso, G.; Maccarini, G. The formability of Aluminum Foam Sandwich panels. *Int. J. Mater. Form.* **2012**, *5*, 243–257. [CrossRef]

21. Olurin, O.B.; Fleck, N.A.; Ashby, M.F. Joining of Aluminum foam with fasteners and adhesives. *J. Mater. Sci.* **2000**, *35*, 1079–1085. [CrossRef]

22. Deshpande, V.S.; Fleck, N.A. High strain rate compressive behaviour of aluminium alloy foams. *Int. J. Imapct Eng.* **2000**, *24*, 277–298. [CrossRef]

23. Jitai, N. Joining of Aluminium Alloy Sheets to Aluminium Alloy Foam Using Metal Glasses. *Metals* **2018**, *8*, 614.

24. Duarte, I.; Krstulović-Opara, L.; Vesenjak, M. Axial crush behaviour of the aluminium alloy in-situ foam filled tubes with very low wall thickness. *Compos. Struct.* **2018**, *192*, 184–192. [CrossRef]

25. Kitazono, K.; Kitajima, A.; Sato, E.; Matsushita, J.; Kuribayashi, K. Solid-state diffusion bonding of closed-cell Aluminum foams. *Mater. Sci. Eng.* **2002**, *A237*, 128–132. [CrossRef]

26. Ubertalli, G.; Ferrais, M.; Bangash, M.K. Joining of AL-6016 to Al-foam using Zn-based joining materials. *Compsites Part A* **2017**, *96*, 122–128. [CrossRef]

27. *ASTM_C393-C393M-06, Standard Test Method for Core Shear Properties of Sandwich Constructions by Beam Flexu-re*; ASTM International: West Conshohocken, PA, America, 2006.

28. *GB/T 1457-2005, Test Method for Climbing Drum Peel Strength of Sandwich Constructions*; Standards Press of China: Beijing, China, 2005.

29. Wang, H.; Yang, D.; He, S.; He, D. Fabrication of open-cell Al foam core sandwich by vibration aided liquid phase bonding method and its mechanical properties. *J. Mater. Sci. Technol.* **2010**, *26*, 423–428. [CrossRef]

30. Wan, L.; Huang, Y.; Lv, S.; Feng, J. Fabrication and interfacial characterization of Aluminum Foam Sandwich via fluxless soldering with surface abrasion. *Compos. Struct.* **2015**, *123*, 366–373. [CrossRef]

31. Kotadia, H.R.; Das, A. Modification of solidification microstructure in hypo- and hyper-eutectic Al–Si alloys under high-intensity ultrasonic irradiation. *Mater. Chem. Phys.* **2011**, *125*, 853–859. [CrossRef]

32. Xu, Z.; Yan, J.; Zhang, B.; Kong, X.; Yang, S. Behaviors of Oxide film at the ultrasonic aided interaction interface of Zn-Al alloy and Al_2O_{3p}/6061Al composites in air. *Mater. Sci. Eng.* **2006**, *A415*, 80–86. [CrossRef]

33. Zhang, S.; Zhao, Y.; Cheng, X.; Chen, G.; Dai, Q. High-energy ultrasonic field effects on the microstructure and mechanical behaviors of A356 alloy. *J. Alloys Compd.* **2009**, *470*, 168–172. [CrossRef]

34. Zheng, D.; Chen, R.; Ma, T.; Ding, H.; Su, Y.; Guo, J.; Fu, H. Microstructure modification and mechanical performances enhancement of Ti44Al6Nb1 Cr alloy by ultrasound treatment. *J. Alloys Compd.* **2017**, *710*, 409–417. [CrossRef]

35. Harte, A.M.; Fleck, N.A.; Ashby, M.F. The fatigue strength of sandwich beams with an aluminium alloy foam core. *Int. J. Fatigue* **2001**, *23*, 499–507. [CrossRef]

36. Burman, M.; Zenkert, D. Fatigue of foam core sandwich beams—1: undamaged specimens. *Int. J. Fatigue* **1997**, *19*, 551–561. [CrossRef]

37. Linul, E.; Serban, D.A.; Marsavina, L.; Kovacik, J. Low-cycle fatigue behaviour of ductile closed-cell aluminium alloy foams. *Fatigue Fract. Eng. Mater. Struct.* **2017**, *40*, 597–604. [CrossRef]

38. Chen, C.; Harte, A.M.; Fleck, N.A. The plastic collapse of sandwich beams with a metallic foam. *Int. J. Mech. Sci.* **2001**, *43*, 1483–1506. [CrossRef]

39. Duart, I.; Teixeir-Dias, F.; Graça, A.; Ferreira, A.J.M. Failure Modes and Influence of the Quasi-static Deformation Rate on the Mechanical Behavior of Sandwich Panels with Aluminum Foam Cores. *Mech. Adv. Mater. Struct.* **2010**, *17*, 335–342. [CrossRef]

Article

Research on Fatigue Strength for Weld Structure Details of Deck with U-rib and Diaphragm in Orthotropic Steel Bridge Deck

Yixin Chen *, Pengmin Lv and Datao Li

Key Laboratory of Road Construction Technology & Equipment, Ministry of Education, Chang'an University, Xi'an 710064, China; lpmin@chd.edu.cn (P.L.); ldt1688@chd.edu.cn (D.L.)
* Correspondence: chenyx@chd.edu.cn; Tel.:+86-29-82334586

Received: 3 April 2019; Accepted: 23 April 2019; Published: 26 April 2019

Abstract: The orthotropic steel bridge deck weld structure would easily cause fatigue cracking under the repeated action of vehicle load. This paper took the steel box girder in a bridge as a research object, researched the mechanical properties of the steel plate and the microstructure of the welded joint, then designed the fatigue specimens of the deck plate and did the fatigue test. The $\Delta\sigma$-N curves and stress amplitudes of the weld details of the deck plate with U-rib and diaphragm under different probabilities of survival were obtained. After extended the $\Delta\sigma$-N curves to the long life range, the fatigue damage calculation equation of the detail was proposed, and the cut-off limit under the 50% and 97.7% probability of survival were 81.50 MPa and 53.11 MPa, respectively. Based on the actual vehicle load spectrum and simplified finite element model of the steel box girder section, the stress amplitude of the details of the weld joint was calculated. The calculation result shows that the maximum stress amplitude of the concerned point was 38.29 MPa, less than the cut-off limit. It means that the fatigue strength of the details of the weld joint meet the requirement of the fatigue design.

Keywords: orthotropic steel bridge deck; weld joint; fatigue design curve; fatigue strength; fatigue test

1. Introduction

Due to the advantages such as a light weight, guarantee of high quality, convenience for in-situ assembling process, etc., the orthotropic steel decks have become very frequently used in the construction of bridges [1–3]. The orthotropic steel bridge deck is a steel deck system with longitudinal ribs and floor beams to support and resist the applied loads such as vehicles and the dead load, and transfer these loads to the main bridge system, which has different types of structures and stiffness values in the longitudinal and transverse directions [4,5]. Much study has shown that the fatigue failure was one of the major factors in the failure of steel bridges [6–9], and which is liable to fatigue cracking once placed in service [10]. Fatigue cracks originated in welded details of the orthotropic steel bridge deck, which greatly impacted on the traffic safety and limited the service life, especially in deck-to-rib joints [11,12]. Under the action of a moving wheel, the damage on the deck plate can highly increase the displacement and stress level of the weld detail because of the changes in its structural behaviors and response generated from the longitudinal ribs against moving loads in the longitudinal direction [13]. It was local stress concentration caused by interaction between out-of-plane motion of U-ribs to adjacent cross-beams and in-plane distortion of the cross-beam that induces the secondary stresses, which may initiate fatigue cracks inside the cross-beams [14]. In addition, there are local stress concentrations on the components caused by residual stresses and weld defects in the weld details, and the geometric abrupt changes at the weld defects and the initial defects of the weld details, which will accelerate the local fatigue crack initiation and expansion [15,16]. Connor and Fisher researched on fatigue resistance of the weld details, and presented a procedure to calculate the stresses at the rib-to-diaphragm joint [17]. Maljaars et al. developed a linear elastic fracture mechanics model for the typical fatigue crack that observed at the root of the weld between the deck plate and stringer [18]. Xiao et al. used the finite element software to study stress analyses and fatigue evaluation of rib-to-deck joints in steel orthotropic decks [19]. Kainuma et al. investigated the fatigue behaviors of the rib-to-deck welded joints and the results showed that the root gap shape and penetration rate have an impact on the root cracking direction and fatigue life [20,21]. Heng et al. evaluated the fatigue performance of rib to deck welded joints in orthotropic steel decks with thickened edge U-ribs, and compared the fatigue performance with the specimens that have conventional U-ribs [22]. Yu et al. did the fatigue test on the typical weld structural details of a steel bridge deck, and the results showed that the cracks tend to occur at the weld locations of the ends and longitudinal ribs and in the bridge deck plate outside the longitudinal ribs [23]. Luo et al. evaluated the fatigue strength of rib-to-deck weld joint in orthotropic steel deck by an average strain energy density method, and derived a W-N curve for the fatigue evaluations [24]. The actual engineering inspection showed that the roof and stiffener welds are most likely to fatigue crack when they are located at the intersection of the diaphragm [25].

However, the stress distribution of the deck plate at its weld joints with U-rib and diaphragm is complicated. The crack is easily initiated in this structural detail, but there are relatively few studies on fatigue behaviors of this detail. In order to make sure the bridge can be working safely, it is necessary to deeply research the fatigue behaviors of the structural detail.

In this paper, the research on the mechanical properties and the microstructure of the steel plate at the welded joint with U-rib and diaphragm were carried out, and then the specimens of the deck plate were designed for the fatigue test. According to the results, the $\Delta\sigma$-N curve and the fatigue damage calculation equation that meets the requirement of the fatigue design of the detail are proposed. This paper also established a simplified finite element model of the steel box girder section, and calculated the stress amplitude of the details of the weld joint using the actual vehicle load spectrum. The calculation result shows that the maximum stress amplitude of the concerned point was less than the cut-off limit.

2. Fatigue Test for Weld Structure of U-Rib and Deck Plate

In order to study the mechanical properties of Q345qD steel plates, a tensile test was carried out using a SD-500 fatigue test machine. Three specimens were designed according to International Standard ISO 6892-1 [26].

According to the standards TB 10002.2-2005 [27] and GB/T 10045-2001 [28], the chemical compositions of Q345qD and the welding wire E501T-1 are shown in Table 1. The test results of the yield strength and the tensile strength are shown in Table 2.

Table 1. The chemical compositions of Q345qD and the welding wire.

Designation	Element (%)					
	C	Si	Mn	P	S	Als
Q345qD	≤0.18	≤0.60	1.1–1.6	≤0.025	≤0.025	≥0.015
E501T-1	≤0.18	≤0.90	≤1.75	≤0.03	≤0.03	-

Table 2. The mechanics performance of Q345qD and the welding wire.

Designation	Specimen Number	Yield Strength/MPa	Tensile Strength/MPa
Q345qD	1	331.3	507.5
	2	342.6	517.3
	3	344.5	509.9
	Mean value	339.5	511.6
E501T-1	1	495	575

2.1. Microstructure Analysis of Welded Joint

The microstructure of the welded joint is different in base metal, weld metal, and the heat affected zone. The local weld joint of the sample was dissected and the sample was corroded in 4% nitric acid alcohol solution for 10 s.

Figure 1a shows the microstructures of the steel specimen, the steel specimen had a ferrite and pearlite banded structure. Figure 1b shows the microstructures of the welded joint, the ferrite was distributed along the columnar crystal on the matrix of sorbite. When the weld was cooled down, the molten metal in the weld was crystallized along the direction of the thermal diffusion, and the columnar crystal was obtained. The eutectoid ferrite separated out along the columnar crystal first, and this ferrite was slightly overheated as the temperature was higher and the cooling rate was slightly faster. During the subsequent cooling process, the austenite turned into acicular distribution of sorbite as greater under-cooling. Figure 1c shows the microstructures of the heat affected zone, and this zone was mainly sorbite and reticular or an acicular distribution of ferrite on the matrix of the Widmanstätten structure.

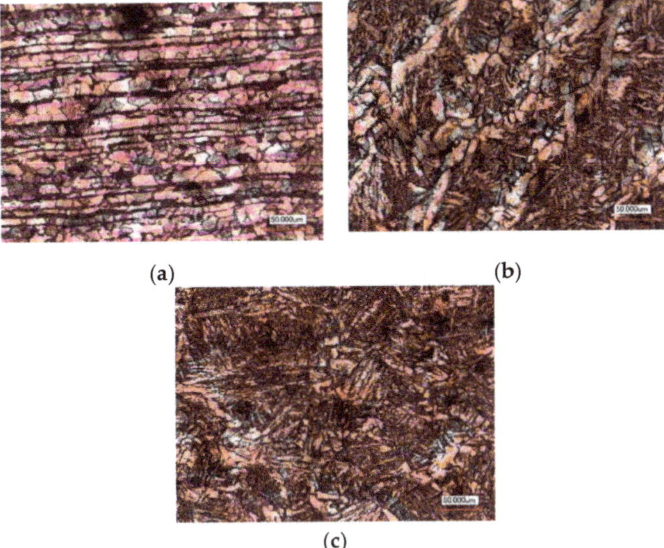

Figure 1. Microstructures of a steel weld joint: (**a**) Steel specimen; (**b**) weld metal; and (**c**) the heat affected zone.

Due to the uneven distribution of the microstructure in the weld area, different degrees of grain coarsening appeared in the fusion zone, which is the weak area of the whole welded joint. The property of the fusion zone was also uneven, which affects the mechanical properties, corrosion resistance, and fatigue resistance of the joint.

2.2. Fatigue Test and Result Analysis

The fatigue specimens were designed and fabricated via V-groove welding. The welding procedure of the specimen was the same as that in the real bridge, which adopted the carbon dioxide arc welding method and used a six-pass welding process. The specific welding parameters are shown in Table 3.

Table 3. Welding parameters.

Electrode Diameter (mm)	Welding Current (A)	Arc Voltage (V)	Welding Speed (cm/min)
Φ1.2	180	24	20

The fatigue test was done by a SD-500 fatigue test machine (produced by Sinotest Equipment Co. ltd, Changchun, China). The stress ratio was set as $R = 0.05$ in the fatigue test. Eleven groups of the fatigue test results were shown in Table 4. Among these eleven specimens, the surface of the specimen sy-2-1-2 and sy-2-1-6 were polished and smoothed to research the influence of surface roughness on fatigue life.

Table 4. Fatigue test results of eleven specimens.

Specimens Number	Stress Amplitudes $\Delta\sigma$/MPa	Number of Cycles N	$\lg\Delta\sigma$	$\lg N$
SY-2-1-1	332.4	152,168	2.522	5.182
SY-2-1-2	272.4	820,188	2.435	5.914
SY-2-1-3	212.5	1,031,746	2.327	6.014
SY-2-1-4	164.5	1,513,565	2.216	6.180
SY-2-1-5	272.4	365,118	2.435	5.562
SY-2-1-6	272.4	1,354,662	2.435	6.132
SY-2-1-7	272.4	290,476	2.435	5.463
SY-2-1-8	272.4	376,123	2.435	5.575
SY-2-1-9	272.4	217,132	2.435	5.337
SY-2-1-10	272.4	677,876	2.435	5.831
SY-2-1-11	176.5	1,419,414	2.247	6.152

According to the data shown in Table 4, the $\Delta\sigma$-N curve equation at 50% probability of survival of the details of the structure was fitted as shown in Equation (1):

$$\lg N = 14.965 - 3.876\lg\Delta\sigma, \tag{1}$$

where, $\Delta\sigma$ is the stress amplitude, N is the number of cycles corresponding to failure. When $N = 2 \times 10^6$, $\Delta\sigma_1$ was equal to 171.92 MPa.

$\Delta\sigma$-N curve equations and stress amplitudes for different probabilities of survival of the structure details were calculated based on Equation (1), as shown in Table 5. The $\Delta\sigma$-N curve equations and stress amplitudes for different probabilities are shown in Figure 2.

Table 5. $\Delta\sigma$-N curve equations and stress amplitude for different failure probabilities.

Probabilities of Survival (%)	$\Delta\sigma$-N Curve Equations	Stress Amplitude $\Delta\sigma$/MPa ($N = 2 \times 10^6$)
60	$\lg N = 14.875 - 3.876\lg\Delta\sigma$	162.95
70	$\lg N = 14.778 - 3.876\lg\Delta\sigma$	153.81
80	$\lg N = 14.663 - 3.876\lg\Delta\sigma$	143.64
90	$\lg N = 14.504 - 3.876\lg\Delta\sigma$	130.74
97.7	$\lg N = 14.244 - 3.876\lg\Delta\sigma$	112.08
99	$\lg N = 14.126 - 3.876\lg\Delta\sigma$	104.44

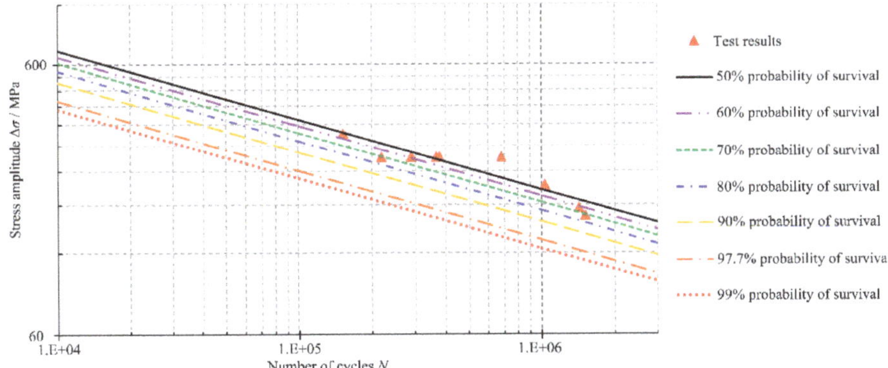

Figure 2. The curves of Δσ-N.

The crack propagation has been monitored by electron microscopy, as shown in Figure 3. According to the fatigue test, the initial crack occurred at the base metal of the bridge deck surface at its welded joint with the diaphragm and U-rib, and then extended lengthwise along the U-rib. The crack location, extension direction, and local view are shown in Figure 3. The crack occurred at the top surface of the bridge deck and was affected by the weld, so the fatigue strength of the details was lower than the base metal, but higher than the weld.

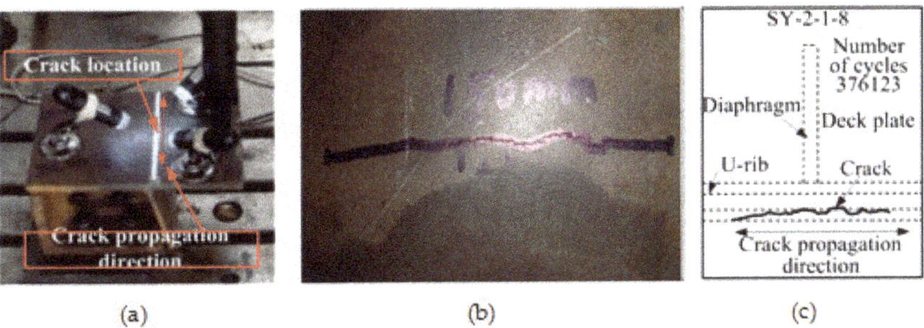

(a)　　　(b)　　　(c)

Figure 3. The crack of the specimen SY-2-1-8: (**a**) Fatigue test; (**b**) crack of the specimen; and (**c**) the position of the crack.

Named the number of cycles when the initial crack occurred as N_1, then compared N_1 with the number of cycles N when the specimen was fractured, and the proportion of initiation life could be obtained, as shown in Table 6.

According to the data shown in Table 6, the Δσ-N_1 curve equation at 50% probability of survival of the details of the structure was fitted as shown in Equation (2):

$$\lg N_1 = 15.717 - 4.219 \lg \Delta\sigma, \tag{2}$$

when $N = 2 \times 10^6$, $\Delta\sigma_2$ was equal to 170.47 MPa.

The Δσ-N_1 curve equation at 97.7% probability of survival of the details of the structure was fitted as shown in Equation (3):

$$\lg N_1 = 14.939 - 4.219 \lg \Delta\sigma, \tag{3}$$

when $N = 2 \times 10^6$, $\Delta\sigma_2$ was equal to 111.53 MPa.

From Table 6 and Figure 4, it could be easily seen that the initiation life of the crack in the weld detail was longer than the crack propagation life.

Table 6. The proportion of initiation life for each specimen.

Specimen Number	Stress Amplitude $\Delta\sigma$/MPa	Number of Cycles when Crack Occurred N_1	Number of Cycles N	Proportion of Initiation Life
SY-2-1-1	332.37	113,210	152,168	74.40%
SY-2-1-3	212.47	964,524	1,031,746	93.48%
SY-2-1-4	164.51	1,440,870	1,513,565	95.20%
SY-2-1-5	272.42	323,484	365,118	88.60%
SY-2-1-7	272.42	230,748	290,476	79.44%
SY-2-1-8	272.42	351,444	376,123	93.44%
SY-2-1-9	272.42	171,784	217,132	79.12%
SY-2-1-10	272.42	592,622	677,876	87.42%
SY-2-1-11	176.50	1,207,128	1,419,414	85.04%

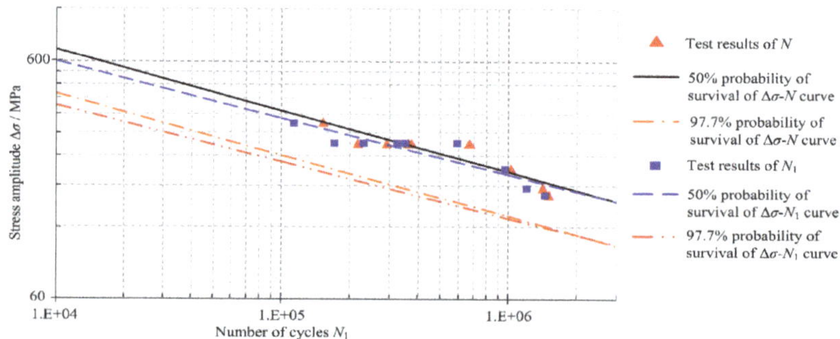

Figure 4. The comparison between the curves $\Delta\sigma$-N and $\Delta\sigma$-N_1.

2.3. The Effect of Surface Roughness on Fatigue Life

According to the principle of fracture mechanics, the larger the surface roughness, the larger the stress concentration coefficient, and leads to a worse fatigue performance.

In order to study the effect of surface roughness on the fatigue life of weld structure of U-rib and deck plate, the surface of the specimen sy-2-1-2 and sy-2-1-6 were smoothed, and the comparison was conducted during the fatigue test. The test results showed that the fatigue life would be improved significantly when the surface of the specimen was polished. The crack location, extension direction, and local view are shown in Figure 5. The number of cycles of the specimen sy-2-1-2 and sy-2-1-6 were calculated by Equation (1) to obtain the theoretical number of cycles. The comparison between the theoretical number of cycles and the test number of cycles is shown in Table 7. The life expectancy rate of the specimen sy-2-1-2 was about 138.2%, and the life-expectancy rate of the specimen sy-2-1-6 was 293.4%.

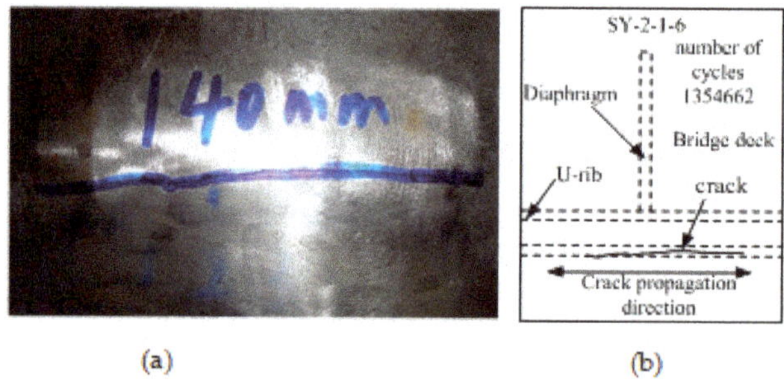

Figure 5. The crack of the specimen SY-2-1-6: (**a**) Crack of the specimen; and (**b**) position of the crack.

Table 7. The comparison between theoretical number of cycles and test number of cycles.

Specimen Number	Theoretical Number of Cycles	Test Number of Cycles	Growth Rate/%
SY-2-1-2	344,372	820,188	138.2%
SY-2-1-6	344,372	1,354,662	293.4%

3. Stress Spectrum Calculation

The fatigue life of the steel bridge deck is related to the stress amplitude and the number of cycles under the action of vehicle load. The calculation of stress amplitude spectrum is important for the fatigue design of steel bridge deck. In order to analyze the stress condition of the deck plate at its weld joint with U-rib and diaphragm, the simplified finite element model of steel bridge was established. The vehicle load spectrum was calculated based on the actual vehicle survey data, and the stress-time history of the concerned point under vehicle load was calculated, then the stress spectrums were obtained by the rain flow counting method.

3.1. Loading Method

According to the actual vehicle survey data, the load frequency spectrum of vehicle load for a certain bridge was obtained. The passing vehicles were divided into seven types on the basis of vehicle type, number of axles, wheelbase, and vehicle weight, as shown in Table 8. The values of lane distribution coefficient were shown in Tables 9 and 10 [29]. The loading method in this paper is single lane loading.

Table 8. Vehicle load frequency spectrum for the certain bridge (96 h).

Vehicle Model	Number of Axles	Illustrations Axle Load/kN, Wheelbase/mm	Total Weight/kN	Number of Vehicles	Ratio of Total Traffic
V1	2	Axle loads: 46, 89; Wheelbase: 4 500	137	19,078	14.74%
V2	3	Axle loads: 46, 54, 122; Wheelbases: 1 950, 5 750	222	2778	2.15%
V3	3	Axle loads: 59, 116, 111; Wheelbases: 4 700, 1 400	318	916	0.71%
V4	4	Axle loads: 65, 72, 91, 91; Wheelbases: 2 100, 4 850, 1 350	317	5342	4.13%
V5	5	Axle loads: 46, 74, 109, 80, 80; Wheelbases: 2 600, 2 000, 10 400, 1 300	375	163	0.13%
V6	5	Axle loads: 60, 105, 73, 73, 73; Wheelbases: 3 600, 6 850, 1 300, 1 300	384	1682	1.30%
V7	6	Axle loads: 45, 73, 110, 80, 80, 80; Wheelbases: 2 550, 2 050, 7 000, 1 300, 1 300	467	18,009	13.91%
Total				47,968	37.06%

Note: The front axle of vehicle had two tires, and the contact area between the wheel and the ground on each side was 0.3 m × 0.2 m. The other axles had four tires, and the contact area on each side was 0.6 m × 0.2 m. The wheel-track of the vehicle was 1.8 m.

Table 9. Lane distribution coefficient of two-axle vehicles.

Vehicle Weight	Vehicle Model	Inside Lane	Center Lane	Outside Lane
0–2 t	Cars, light buses	0.699	0.248	0.054
2–8 t	Bus Two axle truck I	0.244	0.372	0.384
>8 t	Two-axle bus Two axle truck II	0.042	0.650	0.309

Table 10. Lane distribution coefficient of multi-axle vehicles.

Number of Axles	Daily Average Number of Vehicles	Inside Lane	Center Lane	Outside Lane
3	1171	0.076	0.435	0.489
4	3384	0.043	0.310	0.648
5	1284	0.067	0.401	0.533
6	543	0.091	0.420	0.488

3.2. FE Model and Stress Calculation of Steel Bridge Structure

3.2.1. Simplified Finite Element Model of Steel Bridge Structure

The target bridge is designed for six lane highways, which has a main bridge length of 140,500 mm and a main span of 81,800 mm. The length of each side span is 22,900 and 35,800 mm, respectively. The bridge structure layout is shown in Figure 6.

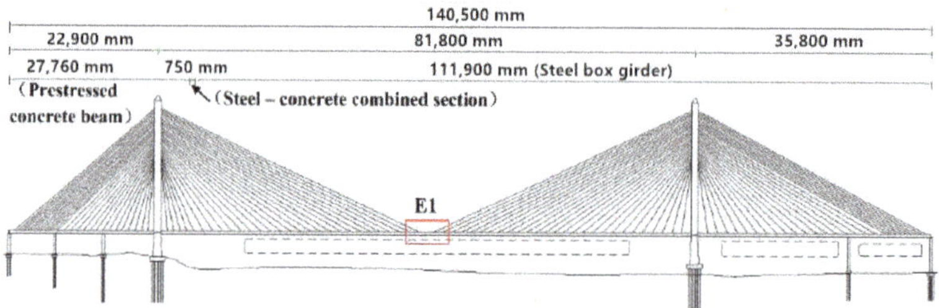

Figure 6. The bridge structure layout.

In this paper, the steel box girder section with the maximum stress variation under the vehicle load was selected as the research object, and named E1. The length of this steel girder section was 15 m, the thickness of the deck plate was 16 mm, the cross section of the longitudinal rib was 300 mm (top width) × 180 mm (bottom width) × 300 mm (height), and the thickness of the rib was 8 mm. The space between two U-ribs was 0.6 m and the space between two diaphragms was 3.75 m. The thickness of the diaphragms was 12 mm and 14 mm, respectively.

Due to the centrosymmetric structure of the steel box girder section, the one-fourth finite element model was established as shown in Figure 7. Based on the St. Venant principle, the simplified finite element model of the steel box girder section was established, as shown in Figure 8. The longitudinal length of the model was 11.25 m (3.75 m × 3 m) and the transverse length of the model was 4.8 m (0.6 m × 8 m). The elastic modulus of steel plate was 210 GPa and the Poisson's ratio was 0.3.

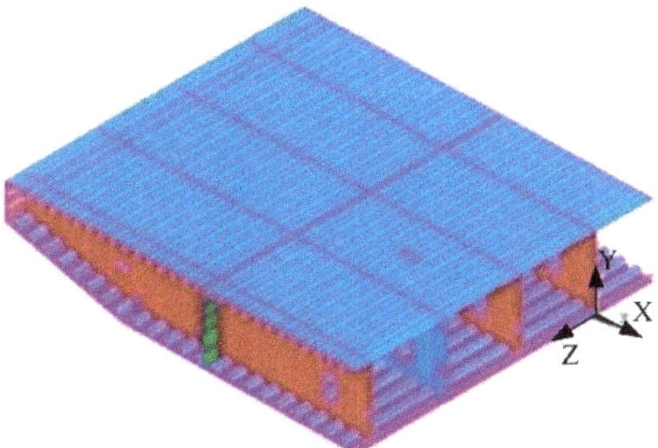

Figure 7. One-fourth finite element model of steel box girder section.

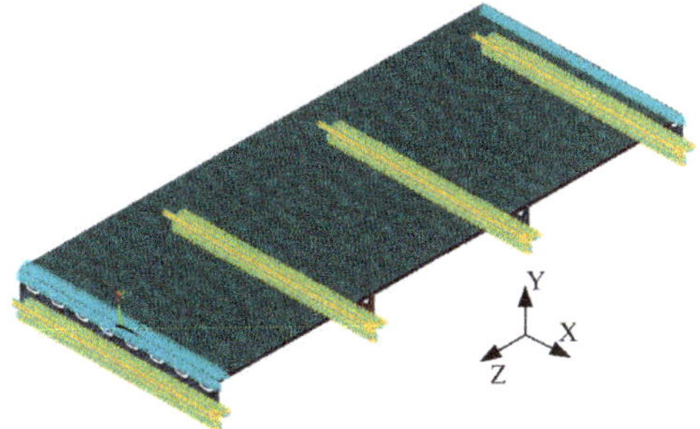

Figure 8. Simplified finite element model of steel box girder section.

Two assumptions were made in the finite element analysis as follows: (1) The components are made of homogeneous, continuous, and isotropic pure elastic materials, and (2) the self-weight and damping of the steel bridge deck are not considered.

Since the box girder webs could restrain the vertical deflection of the deck plate effectively, as shown in Figure 8, along the longitudinal direction (Z-axis) of the girder section, the vertical translation (along Y-axis) of one end of the section was constrained, and the vertical translation and the lateral displacement (along X-axis) of the other end were constrained. All six degrees of freedom of the diaphragms were fixed at the bottom.

3.2.2. Determination of the Concerned Point

The concerned point was located on the top surface of the deck plate at its weld joints with U-rib and diaphragm, as shown in Figure 9. In this position, the U-rib crossed through the perforated diaphragm and connected with the diaphragm by filled welds, and the stress concentration would exist at the perforated position of the diaphragm and the end of the welds. This position would be affected by the in-plane bending stress and the out-of-plane bending stress, which were caused by the deflection of the U-rib. Therefore, the fatigue cracks were easily generated at this concerned point of the deck plate under the action of the changing load of the vehicle.

When the vehicle load acting on the orthotropic steel bridge deck, there was a large stress value in the position of the load and its vicinity, but the stress value away from the load position was very small and the influence distance was much smaller than the vehicle wheelbase. Therefore, when analyzing the worst loading position of the bridge deck, it could use the dual-wheel load of the vehicle [16]. In order to get the worst loading position of the concerned point, take the vehicle model V2 in Table 9 as an example, the stress-time history curve was calculated based on the axle weight of one side of rear axle. The stress-time history curve and the worst loading position of the concerned point are shown in Figures 10 and 11, respectively.

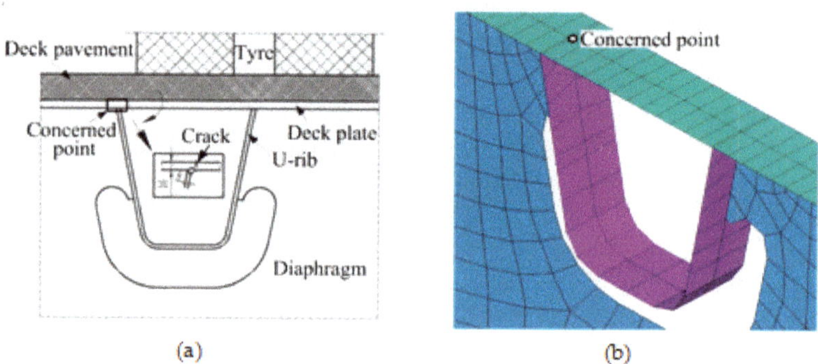

Figure 9. The position of the concerned point: (**a**) Position of the concerned point; and (**b**) the concerned point in the model.

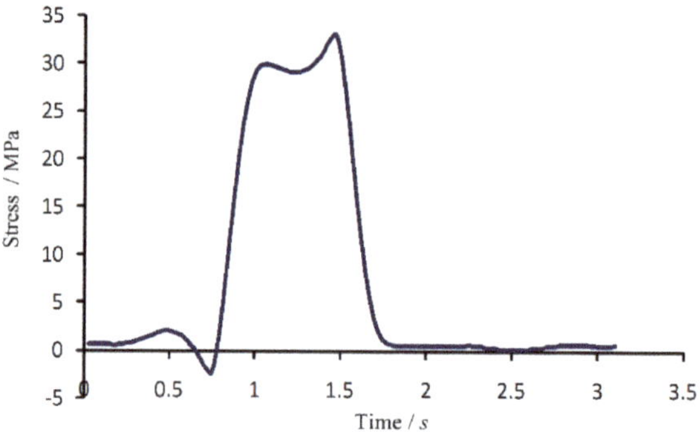

Figure 10. The stress-time curve of concerned point.

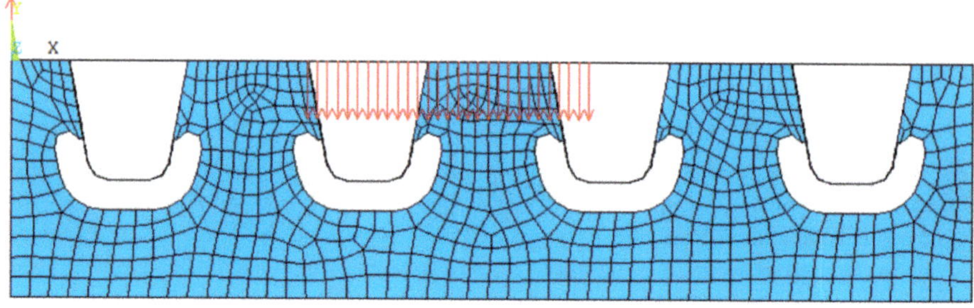

Figure 11. The worst loading position of the concerned point.

3.2.3. The Calculation of Stress Frequency Spectrum

The stress calculated by elastic theory should be included in the impact effect of the vehicle load. According to AASHTO LRFD Bridge Design Specifications [30], the impact coefficient was adopted to be 0.15 in this paper.

Due to the influence of the deck pavement on the contact area between the wheel and the ground, the wheel load was extended to the bridge deck along the 45° direction during the loading process.

Thus, the contact area of the front wheel should be modified to 0.41 m × 0.31 m and the modified contact areas of the remaining wheels were 0.71 m × 0.31 m.

Based on the vehicle load spectrum shown in Table 8, a command stream that let the front axle of the vehicle travel from one end of the bridge deck until the rear axle leaves the other end of the bridge in turn were written by ANSYS APDL. Each load step was recorded to obtain the stress-time history of the concerned point, as shown in Figure 12.

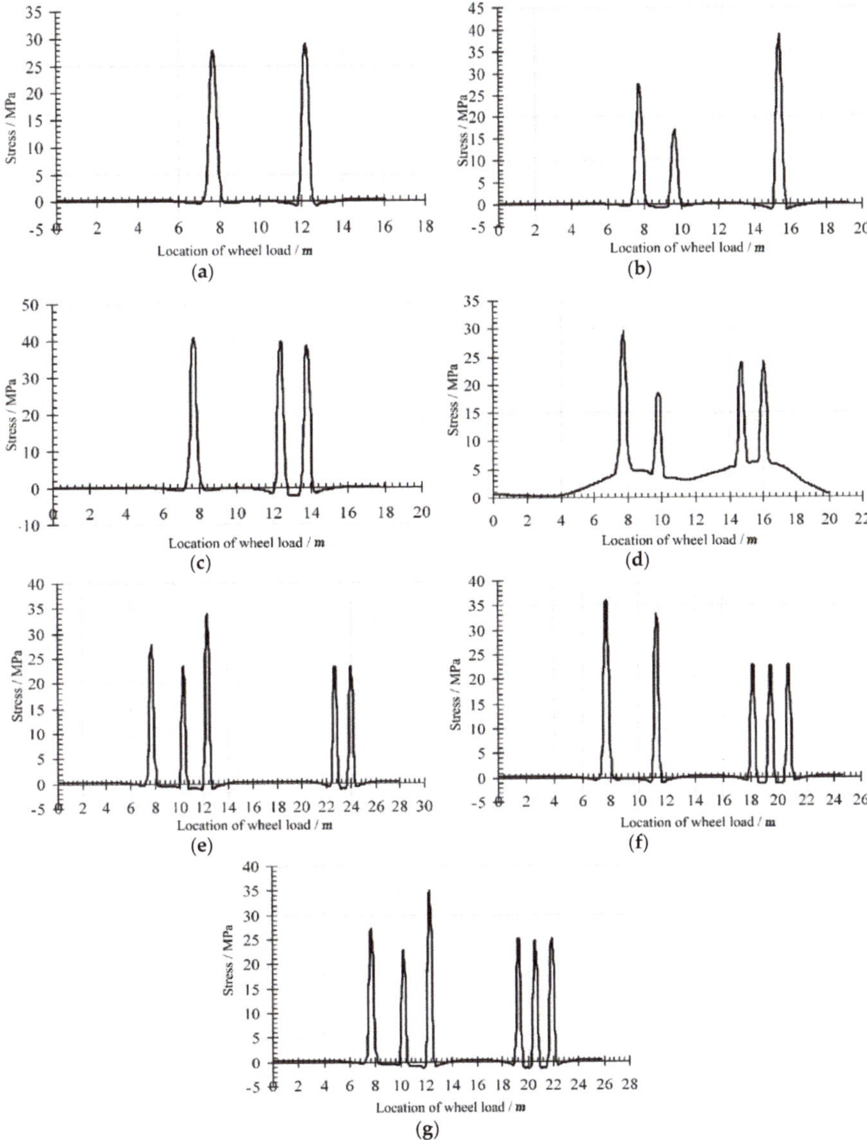

Figure 12. The stress-time curve of concerned point by various vehicles load: (**a**) Vehicle model V1; (**b**) Vehicle model V2; (**c**) Vehicle model V3; (**d**) Vehicle model V4; (**e**) Vehicle model V5; (**f**) Vehicle model V6; and (**g**) Vehicle model V7.

According to the actual vehicle survey data, the stress history of the concerned point was analyzed by the rain flow counting method, and the maximum stress amplitude was 38.29 MPa. The stress amplitude range (0–40 MPa) was divided into eight levels equally [31], and the corresponding number of cycles of the concerned point in one year of each stress level was calculated, as shown in Table 11.

Table 11. Stress amplitude and the corresponding number of cycles (one year).

Stress Level	Stress Amplitude/MPa	Number of Cycles
1	0–5	498,915
2	5–10	0
3	10–15	838,642
4	15–20	183,148
5	20–25	1,139,528
6	25–30	997,920
7	30–35	339,727
8	35–40	78,374
Total		4,076,254

4. Fatigue Life Assessment of the Concerned Point

In order to evaluate the fatigue life of the details of the deck plate at the weld joint with U-rib and diaphragm, the $\Delta\sigma$-N curve obtained from the fatigue test was extended to the long life range according to the Eurocode 3 [32], and then the $\Delta\sigma$-N curve that met the requirement of the fatigue design of the details could be obtained, as shown in Figure 13.

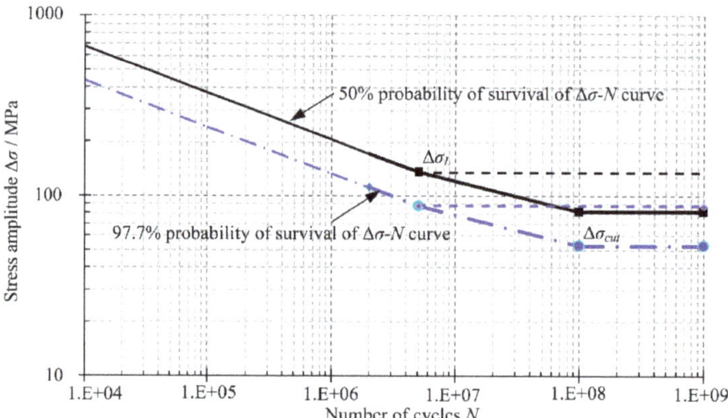

Figure 13. The extended $\Delta\sigma$-N curves.

According to the Eurocode 3, the detail category used to designate a particular fatigue strength curve corresponded to the reference value (in N/mm^2) of the fatigue strength at two million cycles, the constant amplitude fatigue limit corresponded to the fatigue strength for five million cycles, and the cut-off limit corresponded to the fatigue strength for 100 million cycles. According to the test result, when the probability of survival was 50%, the constant amplitude fatigue limit ($N = 5 \times 10^6$) of the detail of the weld joint was $\Delta\sigma_{L50\%} = 135.71$ MPa, and the cut-off limit ($N = 10^8$) was $\Delta\sigma_{cut50\%} = 81.50$ MPa. The maximum stress amplitude of the detail was 38.29 MPa, less than the cut-off limit. When the probability of survival was 97.7%, the constant amplitude fatigue limit ($N = 5 \times 10^6$) of the detail was $\Delta\sigma_{L97.7\%} = 88.43$ MPa, and the cut-off limit ($N = 10^8$) was $\Delta\sigma_{cut97.7\%} = 53.11$ MPa.

The maximum stress amplitude of the detail was 38.29 MPa, less than the cut-off limit. As the maximum stress amplitude of the detail was less than the cut-off limit, it was considered that there was no fatigue damage of the details of the weld joint.

If the maximum stress amplitude of the detail was more than the constant amplitude fatigue limit, combining with the Miner's Rule and Eurocode 3 [32], the amount of fatigue damage of the weld detail of the structure at 50% probability of survival could be calculated as shown in Equation (4), and at 97.7% probability of survival could be calculated as shown in Equation (5):

$$
\begin{cases}
D_i = n_i/N_i = n_i/(5 \times 10^6)\left(\frac{135.71}{\Delta\sigma_i}\right)^{3.876}, & 135.71 \le \Delta\sigma_i \\
D_i = n_i/N_i = n_i/(5 \times 10^6)\left(\frac{135.71}{\Delta\sigma_i}\right)^{5.876}, & 81.50 \le \Delta\sigma_i < 135.71 \\
D_i = n_i/N_i = 0, & \Delta\sigma_i \le 81.50
\end{cases}
\tag{4}
$$

$$
\begin{cases}
D_i = n_i/N_i = n_i/(5 \times 10^6)\left(\frac{88.43}{\Delta\sigma_i}\right)^{3.876}, & 88.43 \le \Delta\sigma_i \\
D_i = n_i/N_i = n_i/(5 \times 10^6)\left(\frac{88.43}{\Delta\sigma_i}\right)^{5.876}, & 53.11 \le \Delta\sigma_i < 88.43 \\
D_i = n_i/N_i = 0, & \Delta\sigma_i \le 53.11
\end{cases}
\tag{5}
$$

The total amount of fatigue damage of the details of the structure could also be calculated as shown in Equation (6) [33]:

$$
D = \sum_i D_i = \sum \frac{n_i}{N_i},
\tag{6}
$$

where D is the total amount of fatigue damage, $\Delta\sigma_i$ is the i-th stress amplitude, N_i is the total number of cycles of the i-th stress amplitude, n_i is the number of cycles of the i-th stress amplitude, and $i = 1, 2, 3, 4, 5, \ldots$.

5. Conclusions

This paper focused on the fatigue strength of the weld joint of the deck plate at its weld joint with U-rib and diaphragm in an orthotropic steel bridge deck, researched the mechanical properties of the steel plate and the microstructure of the welded joint, then designed the fatigue specimens of the deck plate and did the fatigue test. The fatigue test results showed that the fatigue life would be improved significantly when the surface of the specimen was polished and smoothed. The test results also showed that the initiation life of the crack in the weld detail was longer than the crack propagation life. The $\Delta\sigma$-N curves and stress amplitudes of the deck plate at its weld joint with U rib and diaphragm under a different probability of survival were obtained. After extending the $\Delta\sigma$-N curves to the long life range, the extended $\Delta\sigma$-N curve and the amount of fatigue damage calculation equation that met the requirement of the fatigue design of the detail was proposed, and the cut-off limit under the 50% and 97.7% probability of survival were 81.50 MPa and 53.11 MPa, respectively.

This paper also established a simplified finite element model of the steel box girder section, determined the concerned point where the fatigue cracks easily generated, selected a reasonable loading process and the impact coefficient, and calculated the stress amplitude of the details of the weld joint using the actual vehicle load spectrum. The calculation result showed that the maximum stress amplitude of the concerned point was 38.29 MPa, less than the cut-off limit. It meant that the fatigue strength of the details of the weld joint met the requirement of the fatigue design.

Author Contributions: Methodology, Y.C.; software, Y.C. and D.L.; supervision, P.L.; writing—original draft preparation, Y.C.; writing—review and editing, Y.C. and P.L.

Funding: This research was funded by the Fundamental Research Funds for the Central Universities of Chang'an University, grant number 310825171012 and 300102259302.

References

1. Ya, S.; Yamada, K.; Shikawa, T. Fatigue evaluation of rib-to-deck welded joints of orthotropic steel bridge deck. *J. Bridge Eng.* **2013**, *18*, 492–499. [CrossRef]
2. Oh, C.K.; Hong, K.J.; Bae, D.; Do, H.; Han, T. Analytical and experimental studies on optimal details of orthotropic steel decks for long span bridges. *Int. J. Steel Struct.* **2011**, *11*, 227–234. [CrossRef]
3. Fu, Z.; Ji, B.; Zhang, C.; Li, D. Experimental study on the fatigue performance of roof and U-rib welds of orthotropic steel bridge decks. *KSCE J Civ. Eng.* **2018**, *22*, 270–278. [CrossRef]
4. Freitas, S.T.D.; Kolstein, H.; Bijlaard, F. Fatigue behavior of bonded and sandwich systems for strengthening orthotropic bridge decks. *Compos. Struct.* **2013**, *97*, 117–128. [CrossRef]
5. Kainuma, S.; Ahn, J.H.; Jeong, Y.S.; Sugiyama, H.; Iwasaki, M. Evaluation of structural responses on artificial fatigue crack for bulb rib orthotropic deck. *Adv. Struct. Eng.* **2016**, *18*, 1355–1369. [CrossRef]
6. Bocchieri, W.J.; Fisher, J.W. *Williamsburg Bridge Replacement Orthotropic Deck as-Built Fatigue Test*; ATLSS Report; Lehigh University: Bethlehem, PA, USA, 1998.
7. Connor, R.J.; Richards, S.O.; Fisher, J.W. Long-term remote monitoring of prototype orthotropic deck panels on the Bronx-Whitestone Bridge for fatigue evaluation. In *Recent Developments in Bridge Engineering*; Swets &Zeitlinger B.V.: Lisse, The Netherlands, 2003; pp. 257–268.
8. Cullimore, M.S.G.; Webber, D. Analysis of heavy girder bridge fatigue failures. *Eng. Fail. Anal.* **2000**, *7*, 145–168. [CrossRef]
9. Choi, D.H.; Choi, H.Y.; Choi, J.H. The effect of diaphragm inside trough rib on fatigue behavior of trough rib and cross beam connections in orthotropic steel decks. *J. Korean Soc. Steel Constr.* **2000**, *23*, 239–250.
10. Nguyen, H.T.; Chu, Q.T.; Kim, S.E. Fatigue analysis of a pre-fabricated orthotropic steel deck for light-weight vehicles. *J. Constr. Steel Res.* **2011**, *67*, 647–655. [CrossRef]
11. Battista, R.C.; Pfeil, M.S.; Carvalho, E.M.L. Fatigue life estimates for a slender orthotropic steel deck. *J. Constr. Steel Res.* **2008**, *64*, 134–143. [CrossRef]
12. Cui, C.; Zhang, Q.H.; Luo, Y.; Hao, H.; Li, J. Fatigue reliability evaluation of deck-to-rib welded joints in OSD considering stochastic traffic load and welding residual stress. *Int. J. Fatigue* **2018**, *111*, 151–160. [CrossRef]
13. Farreras-Alcover, I.; Chryssanthopoulos, M.K.; Andersen, J.E. Data-based models for fatigue reliability of orthotropic steel bridge decks based on temperature, traffic and strain monitoring. *Int. J. Fatigue* **2017**, *95*, 104–119. [CrossRef]
14. Kainuma, S.; Jeong, Y.S.; Ahn, J.H.; Yamagami, T.; Tsukamoto, S. Behavior and stress of orthotropic deck with bulb rib by surface corrosion. *J. Constr. Steel Res.* **2015**, *113*, 135–145. [CrossRef]
15. Wolchuk, R. Lessons from weld cracks in orthotropic decks on three European bridges. *J. Struct. Eng.* **1990**, *116*, 75–84. [CrossRef]
16. Tsakopoulos, P.A.; Fisher, J.W. Full-scale fatigue tests of steel orthotropic deck panel for the Bronx—Whitestone Bridge rehabilitation. *Bridge Struct.* **2005**, *1*, 55–66. [CrossRef]
17. Connor, R.J.; Fisher, J.W. Consistent approach to calculating stresses for fatigue design of welded rib-to-web connections in steel orthotropic bridge decks. *J. Bridge Eng.* **2006**, *11*, 517–525. [CrossRef]
18. Maljaars, J.; Bonet, E.; Pijpers, R.J.M. Fatigue resistance of the deck plate in steel orthotropic deck structures. *Eng. Fract. Mech.* **2018**, *201*, 214–228. [CrossRef]
19. Xiao, Z.G.; Yamada, K.; Ya, S.; Zhao, X.L. Stress analyses and fatigue evaluation of rib-to-deck joints in steel orthotropic decks. *Int. J. Fatigue* **2008**, *30*, 1387–1397. [CrossRef]
20. Kainuma, S.; Yang, M.Y.; Jeong, Y.S.; Inokuchi, S.; Kawabata, A.; Uchida, D. Experimental investigation for structural parameter effects on fatigue behavior of rib-to-deck welded joints in orthotropic steel decks. *Eng. Fail. Anal.* **2017**, *79*, 520–537. [CrossRef]
21. Kainuma, S.; Yang, M.Y.; Jeong, Y.S.; Inokuchi, S.; Kawabata, A.; Uchida, D. Fatigue Behavior Investigation and Stress Analysis for Rib-to-Deck Welded Joints in Orthotropic Steel Decks. *Int. J. Steel Struct.* **2018**, *18*, 512–527. [CrossRef]
22. Heng, J.; Zheng, K.; Gou, C.; Zhang, Y.; Bao, Y. Fatigue performance of rib-to-deck joints in orthotropic steel decks with thickened edge U-ribs. *J. Bridge Eng.* **2017**, *22*, 04017059. [CrossRef]
23. Yu, B.; Qiu, H.X.; Wang, H.; Guo, T. Experimental research on fatigue behavior and damage development of welded conformation of orthotropic steel bridge deck. *J. Highway Transp. Res. Dev.* **2009**, *26*, 64–69.

24. Luo, P.J.; Zhang, Q.H.; Bao, Y.; Zhou, A.X. Fatigue evaluation of rib-to-deck welded joint using averaged strain energy density method. *Eng. Struct.* **2018**, *177*, 682–694. [CrossRef]

25. Zeng, Z.B. Classification and reasons of typical fatigue cracks in orthotropic steel deck. *Steel Constr.* **2011**, *26*, 9–16.

26. ISO. Metallic materials-tensile testing, Part 1: Method of test at room temperature. *Int. Organiz. Stand.* **2016**, 6892-1.

27. TB 10002.2. *Code for Design on Steel Structure of Railway Bridge*; China Railway Publishing House: Beijing, China, 2005.

28. GB/T 10045-2001. *Carbon Steel Flux Cored Electrodes for Arc Welding*; Standards Press of China: Beijing, China, 2001.

29. Zhou, Y.B. Crack Study and Local Fatigue Analysis of Orthotropic Steel Decks on Bridges. Master's Thesis, Tsinghua University, Beijing, China, 2010.

30. AASHTO. *AASHTO LRFD Bridge Design Specifications*; American Association of State Highway and Transportation Officials: Washington, DC, USA, 2007.

31. Sonsino, C.M. Fatigue testing under variable amplitude loading. *Int. J. Fatigue* **2007**, *29*, 1080–1089. [CrossRef]

32. EN. *Eurocode 3: Design of Steel Structures, Part 1–9: Fatigue Strength of Steel Structures*; European Committee for Standardization: Brussels, Belgium, 2003.

33. Lee, Y.L.; Pan, J.; Hathaway, R.; Barkey, M. *Fatigue Testing and Analysis: Theory and Practice (Vol. 13)*; Butterworth-Heinemann: Burlington, MA, USA, 2005.

Article

Effect of Repetitive Collar Replacement on the Residual Strength and Fatigue Life of Retained Hi-Lok Fastener Pins

David F. Hardy [†] and David L. DuQuesnay *

Department of Mechanical and Aerospace Engineering, Royal military College of Canada, PO Box 17000 Station Forces, Kingston, ON K7K 7B4, Canada; David.Hardy@forces.gc.ca
* Correspondence: duquesnay-d@rmc.ca; Tel.: +613-541-6000 (ext. 6483)
† Currently: DT4 Structural Engineer, Aerospace Telecommunications and Engineering Support Squadron, K0K 3W0 Astra, ON, Canada.

Received: 13 March 2019; Accepted: 9 April 2019; Published: 16 April 2019

Abstract: Hi-Lok fasteners were subjected to multiple collar replacements, and were tested under static loading and constant-amplitude fatigue loading, to determine the effect of repetitive collar replacement on the residual strength and fatigue life of a retained Hi-Lok-type fastener pin. Hi-Lok-type fasteners are typically used in aircraft structural joints, and are loaded mainly in shear. Tests were conducted for clamping force, static shear strength, static tensile strength, and shear fatigue life for collars subjected to five collar replacements. The static shear results showed no decrease in the ultimate shear strength of the fastener pin as a function of collar replacement. Static tensile results showed no decrease in the ultimate tensile strength of the fastener as a function of collar replacement, with failure of the aluminum collar remaining the critical failure mode. Similarly, shear fatigue results showed no decrease in the shear fatigue life of the fastened joint as a result of collar replacement, with fracture of the aluminum substrate remaining the critical failure mode. For static shear, static tension, and shear fatigue tests, estimated clamping force was highly consistent between specimens and no decrease in clamping force was observed as a function of collar replacement.

Keywords: aircraft; fatigue; fastener; reuse

---

1. Introduction

Hi-Lok-type [1] fasteners are used extensively in the aerospace industry, and consist of a threaded fastener pin and collar. The threaded collar includes a wrenching element that shears-off at a fixed torque, enabling rapid installation while ensuring a consistent pre-load for each fastener. The manufacturer's installation instructions recommend retention and reuse of undamaged Hi-Lok fastener pins following collar removal [2]; however, current Royal Canadian Air Force (RCAF) maintenance procedures [3] call for the replacement of an installed aircraft Hi-Lok fastener pin each time the collar is removed.

Despite this current practice, it may be preferable from an aircraft maintenance standpoint to remove and replace the collars without replacing the installed fastener pins in some instances. Specifically, retention of the installed pins may be preferable when accessing fastener holes for Non-Destructive Inspection (NDI) of aircraft structure using techniques such as Eddy Current Surface Scan (ECSS). NDI necessitating scheduled collar removal is often carried out in structural areas where fatigue cracking is a concern. By far, the most commonly used technique to detect fatigue damage in and around aircraft fasteners involves eddy current (EC) inspection. Fhar and Wallace [4] provide a comprehensive discussion of various NDI techniques used to inspect fastener holes in aircraft structures. The most common and reliable method is the bolt hole eddy current (BHEC) that requires the removal of the fastener itself to permit a probe into the faster hole for inspection. However, the removal and re-installation of interference fit fastener pins causes mechanical damage to the fastener hole due to strain and physical abrasion. It is clear that repetitive removal and installation of interference fit fastener pins throughout the life of the aircraft will produce far more mechanical damage than if the initial fastener pin were retained. Thus, ironically, the periodic removal and reinstallation of interference fit fastener pins for NDI of fatigue-prone areas risks exacerbating the existing fatigue concerns at these locations by generating or aggravating existing crack initiations. Although not as powerful as BHEC at detecting cracks, especially cracks within the bore of faster hole, ECSS can still be used to detect surface cracks. The removal of the fastener collar enables detection of cracks at the edge of the fastener hole, rather than only detecting cracks that extend past the edge of the installed collar. It has been shown that fatigue crack growth in aircraft structure under a variety of loading conditions follows an exponential growth rate with usage (cycles or flights) with the majority of fatigue lifetime spent with cracks relatively small, approximately 5 mm or less [5]. Because of the exponential nature of crack growth, the ability to detect fatigue cracks while they remain relatively short significantly increases the allowable time between inspections, considerably reducing manpower requirements and maintenance downtime for the affected aircraft fleet.

The replacement of fastener collars for NDI without the removal of the interference fit fastener pins may therefore help to maximize the fatigue life of the aircraft structure being inspected. An example of a structural area where repetitive collar replacement was considered for use in the RCAF is shown in Figure 1 for Critical Point 16 on the CP140 Aurora [6]. Selected hardware, test specimens, and applied loads for this project were based on the fasteners and structure at this location. Although the fatigue behavior of fasteners and joints in aircraft structure has been studied extensively [5,7], only a few studies have been done on the effect of repeated re-installation of fasteners. Li and Su [8] studied the effect of repeatedly tightening and loosening the nut on titanium fasteners in a carbon fibre composite structure. They found that the clamping load decreased rapidly with repeated loosening-tightening cycles with a drop in clamp load between 66–76% within five repetitions and down to about 45% after 15 repetitions. Clamping load is critical to the fatigue performance of typical fastened aircraft structural joint, with a decrease in clamping load leading to early fatigue failure [7].

Figure 1. CP140 Critical Point 16 inboard engine nacelle skate angle.

Despite the current RCAF practice of replacing fastener pins and collars at each inspection, it is evident that there is a benefit, and it would be convenient, to retain the faster pins in some situations while removing only the collar to allow clear access to the fastener hole for inspection. The effect of repeated replacement of the collar on Hi-Lok pins on the subsequent strength and life of the fastener, and on the clamping load has not yet been reported. The aim of this research was to determine the effect of repetitive collar replacement on the residual clamping force, static strength, and fatigue life of retained Hi-Lok-type fastener pins. Fasteners were tested separately in both tension and shear to determine their behaviour under each primary loading condition. The results of this study will provide some evidence to support the practice of repeated collar replacements on retained Hi-Lok fasteners in aircraft structure where pin removal would be difficult, as is the case for Critical Point 16 on the CP140 Aurora [6].

1.1. Equivalent Interference Fits

Most aerospace applications of Hi-Lok-type fasteners use interference fits between the pins and fastener holes to generate localized compressive stresses that reduce crack growth rates, improving the fatigue life of the component. Critical Point 16 on the CP140 Aurora is taken as a representative example of a fatigue-prone aircraft structure where Hi-Lok-type fasteners are employed. This location calls for 0.245″ to 0.248″ (6.22 mm to 6.30 mm) diameter fastener holes [9,10] in the 7075-T6 aluminum structure to accommodate 0.249″ (6.32 mm) diameter steel fastener pins, resulting in an interference of 0.001″ to 0.004″ (0.025 mm to 0.100 mm) between the pins and the substrate.

In order to generate an equivalent amount of physical interference and resultant mechanical damage in the test specimens for this project, 0.245″ (6.22 mm) diameter reamed fastener holes were used for all aluminum shear test fixtures. Because strength and fatigue requirements resulted in some of the test specimens, such as the tensile test jigs, being manufactured from high-strength steel, the physical interference of fastener holes in the steel test fixtures is reduced to ensure a similar level of mechanical damage to the fastener pin during installation. When determining the appropriate level of interference fit, it is assumed that the mechanical damage to the pin during installation is proportional to its radial strain.

For the aluminum plates in the shear test fixtures, total deformation at the fastener hole, ΔD_{total}, is equal to the difference between the diameters of the fastener pin, D_{pin}, and fastener hole, D_{hole}:

$$\Delta D_{total} = D_{pin} - D_{hole}. \tag{1}$$

Part of this deformation is caused by negative strain (radial compression) of the pin, while the remainder is accommodated by positive strain (radial expansion) of the fastener hole.

The deformation of the pin can be expressed in terms of the deformation of the fastener hole, knowing that the relative strains of these components will be directly proportional to the relative stiffness of the pin and plate materials:

$$\frac{|\varepsilon_{pin}|}{\varepsilon_{hole}} = \frac{E_{plate}}{E_{pin}}, \tag{2a}$$

$$\frac{E_{plate}}{E_{pin}} = \frac{\left(\frac{|\Delta D_{pin}|}{D_{pin}}\right)}{\left(\frac{\Delta D_{hole}}{D_{hole}}\right)}, \tag{2b}$$

$$= \frac{D_{hole}|\Delta D_{pin}|}{D_{pin}\Delta D_{hole}}, \tag{2c}$$

$$|\Delta D_{pin}| = \frac{D_{pin}\Delta D_{hole}}{D_{hole}}\left(\frac{E_{plate}}{E_{pin}}\right), \tag{2d}$$

where $|\varepsilon_{pin}|$ is the absolute radial strain of the pin (by convention, compressive strain will have a negative value),

ε_{hole} is the radial strain of the fastener hole,

E_{pin} is the elastic modulus of the HL51-8-6 fastener pin material, 2.9×10^4 ksi (199 GPa) [11],

E_{plate} is the elastic modulus of the 7075-T6 plate material, 1.03×10^4 ksi (70 GPa) [11],

$|\Delta D_{pin}|$ is the absolute radial deformation of the pin (compressive deformation will have a negative value), and

ΔD_{hole} is the radial deformation of the fastener hole.

Substitute Equation (1) into Equation (2) and solve for the deformation of the pin:

$$|\Delta D_{pin}| = \Delta D_{total}\left(\frac{A}{1+A}\right), \tag{3}$$

where $A = \frac{D_{pin}E_{plate}}{D_{hole}E_{pin}}$.

Calculating deformation of the fastener pin in the aluminum plates:

$$\Delta D_{total} = D_{pin} - D_{hole} = (0.249) - (0.245) = 0.004'' \ (0.100 \text{ mm}), \tag{4a}$$

$$A = \frac{D_{pin}E_{plate}}{D_{hole}E_{pin}} = \frac{(0.249)\left(1.03 \times 10^4\right)}{(0.245)\left(2.90 \times 10^4\right)} = 0.36, \tag{4b}$$

$$|\Delta D_{pin}| = \Delta D_{total}\left(\frac{A}{1+A}\right) = 4.00 \times 10^{-3}\left(\frac{0.36}{1+0.36}\right) = 0.001'' \ (0.025 \text{ mm}). \tag{4c}$$

Because aluminum has a much lower elastic modulus than steel, aluminum test jigs will have a greater amount of deformation relative to the steel fastener pin than steel test jigs, assuming equal dimensions. In order to ensure the same level of mechanical damage for the fastener pins in the steel axial test jig, radial strain (and therefore radial deformation) of the fastener pin is held constant.

To find the interference fit that will generate the same radial deformation of the pin in a high-strength steel fixture, Equation (2) is solved for deformation of the fastener hole:

$$|\Delta D_{pin}| = \frac{D_{pin}\Delta D_{hole}}{D_{hole}}\left(\frac{E_{plate}}{E_{pin}}\right), \tag{5a}$$

$$\Delta D_{hole} = \frac{D_{hole}|\Delta D_{pin}|}{D_{pin}}\left(\frac{E_{pin}}{E_{plate}}\right). \tag{5b}$$

Rearranging Equation (1), the initial diameter of the fastener hole is equal to:

$$D_{hole} = |\Delta D_{pin}| - \Delta D_{total}. \tag{6}$$

Because the initial fastener hole diameter and radial deformation of the fastener hole are interdependent, the equations above are iterated for increasing values of D_{hole} until a solution converges. For a defined absolute pin deformation of $0.0011''$ (0.027 mm):

$D_{hole} = 0.247''$ (6.30 mm).

A $0.247''$ diameter (6.30 mm) reamed fastener hole was therefore used for the high-strength steel test fixtures.

1.2. Measurement of Clamping Force

Clamping force or "preload" is a major consideration for fastened joints. The application of a tensile preload to a fastener pin improves the fatigue life of the pin and the joint by decreasing the effective stress amplitude. Estimation of the clamping force in a fastened joint is therefore an important metric when assessing the impact of repetitive collar replacement on a retained fastener pin.

The tightening torque, T_{on}, of a fastener can be related to the induced clamping force, F, by the following expression described by Eccles [12,13]:

$$T_{on} = \frac{F}{2}\left[\frac{\mu_t d_2}{\cos\beta} + \frac{p}{\pi} + D_e\mu_n\right],$$

(7)

where μ_t is the coefficient of friction for the thread;

d_2 is the basic pitch diameter of the thread;

β is the half-included angle of the threads (30° for UNF threads);

p is the pitch of the threads;

μ_e is the coefficient of friction for the nut face; and

$D_e = \frac{1}{2}(d_i + d_o)$ is the effective bearing diameter of the nut,

where d_i and d_o are the inner and outer bearing diameters of the nut, respectively.

The loosening torque, T_{off}, can similarly be related to the clamping force by the expression:

$$T_{off} = \frac{F}{2}\left[\frac{\mu_t d_2}{\cos\beta} - \frac{p}{\pi} + D_e\mu_n\right]$$

(8)

From these equations, the difference between the tightening and loosening torques can be simplified into the following expression:

$$T_{on} - T_{off} = F\left(\frac{p}{\pi}\right).$$

(9)

Rearranging Equation (8), the clamping force in the joint can be simply related to the difference between the tightening and loosening torques, as follows:

$$F = \frac{\pi}{p}\left(T_{on} - T_{off}\right).$$

(10)

The uncertainty of clamping force estimates calculated using this method varies depending on the method of torque application, but is generally found to fall within ±17% using a precision torque wrench [7].

While Eccles [12,13] recommends the use of an *on–off–on* measurement process to record tightening and loosening torques for the fastener, this process is not practical for Hi-Lok-type collars. Because the wrenching element of the collar torques-off at a pre-determined torque during installation, subsequent tightening of the collar is not easily achieved and is difficult to measure accurately. Moreover, loosening of the collar can only be achieved and accurately measured by damaging means, using specialized Hi-Lok removal tools. For the purposes of this experiment, tightening torque, T_{on}, is therefore assumed to be equal to the peak torque achieved during installation of the collar. Loosening torque, T_{off}, is taken as the peak torque achieved during removal of the collar.

2. Materials and Methods

The fasteners used in this project are the Hi-Lok HL51-8-6 (HL19 equivalent) 0.25″ (6.35 mm) diameter 100° flush shear head cadmium-plated alloy steel fastener pins with HL70-8 aluminum collars. These are the countersunk fastener type used on the CP140 Aurora at Critical Point 16, and are used extensively on CP140 and CC130 aircraft. Moreover, these alloy steel flush shear head fasteners are broadly representative of similar Hi-Lok-type fasteners used across many other RCAF aircraft fleets. This type of fastener is designed and intended to take shear loading. The specified mechanical properties of the fasteners and collars are given in Tables 1 and 2, respectively.

Table 1. Mechanical properties of HL51-8-6 fastener pin [11].

Property	Value (as Specified)	Value (SI Equivalent)
Shaft diameter, D	0.249 ± 0.010″	6.32 ± 0.025 mm
Grip Length, L	0.375 ± 0.005″	9.52 ± 0.127 mm
Thread	0.250-28 UNJF-3A	-
Material	Alloy Steel	-
Surface Treatment	Cadmium	199 GPa
Elastic Modulus	29,000 ksi	
Ultimate Single Shear Strength, P_{su}	≥4650 lb$_f$	≥20.7 kN
Ultimate Tensile Strength, P_{tu}	≥3700 lb$_f$	≥16.5 kN
Ultimate Shear Stress, F_{su}	≥95 ksi	≥654 MPa
Ultimate Tensile Stress, F_{tu}	160–180 ksi	1100–1240 MPa

Table 2. General properties of HL70-8 collar [14].

Property	Value (as Specified)	Value (SI Equivalent)
Outer Diameter, D	0.249 ± 0.010″	6.32 ± 0.025 mm
Retained Length, L (*)	0.375 ± 0.005″	9.52 ± 0.127 mm
Thread	0.250-28 UNJF-3B	-
Material	2024-T6 Aluminum	-
Ultimate Tensile Strength, P_{tu}	≥3000 lb$_f$	13.3 kN
Torque (on installation)	60–80 lb$_f$ - in	6.78–9.04 Nm

* Length after torque-off of wrenching element.

Fasteners were installed and removed using a DTL-100i Digital Torque Wrench and compatible 3/8″ socket, an Allen key, and SAVI-32, -34, and -36 Hi-Lok removal tools.

2.1. Procedures

Unless otherwise shown, ten repetitions of each specimen type were tested to generate statistically significant results. All shear test specimens were made with the same dimensions, consisting of two strips of material measuring 1.50″ × 4.50″ × 0.125″ (38.1 mm × 114 mm × 3.17 mm) with a single fastener installed in the overlap. An edge distance, e = 0.50″ (12.7 mm) was used for all fastener holes. This edge distance is reflective of typical aircraft structure and allows direct comparison with manufacturers' material strength values, which are typically quoted for edge distances of 2D or greater.

2.2. Baseline Static Shear

Fasteners were installed in accordance with [3] into single-fastener lap joints made from AISI Type O1 steel plates and tested monotonically to failure. The test fixture is illustrated in Figure 2. The ultimate shear strength and failure mode were recorded for each specimen.

2.3. Static Shear with Collar Replacement

Following completion of the baseline tests, static shear tests were conducted on fasteners subjected to multiple collar replacements. Fasteners were installed in accordance with [3] into single-fastener interference fit lap joints made from 7075-T6 aluminum sheet. Collars were removed and re-installed in accordance with [3] for a total of five collar replacements. The loosening and tightening torques for all collars were measured using a digital torque wrench to enable calculation of the clamping force of each collar in accordance with [11]. Following the fifth and final collar replacement, the fastener pins were removed from the aluminum plate specimens and installed with a clearance fit in accordance with [3] into heat treated AISI Type O1 steel sheet (Figure 2), which provided the high strength required for shear testing of the fastener pins. In an aircraft inspection scenario, the possibility of damaging the substrate material would be a major concern, but, in our case, damage to the fastener pin was of interest. The pins were carefully pressed out in a single continuous motion and then visually inspected under light microscope (up to 20×). No discernable damage was visible on any fastener shank. The fasteners were then tested monotonically to failure to determine the ultimate shear strength and failure mode for each specimen.

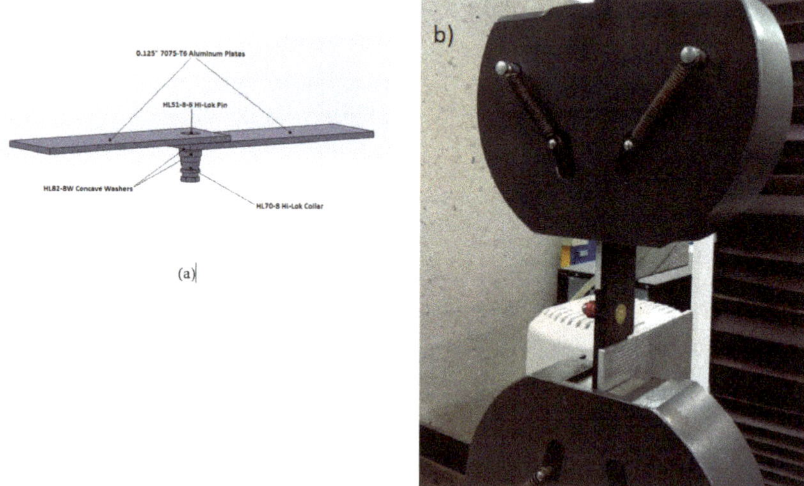

Figure 2. Static shear test assembly (**a**) aluminum joint for repeated collar installation and (**b**) steel shear test fixture installed in test machine.

2.4. Static Tension

Tests were conducted to determine the baseline ultimate tensile strength of the fasteners with no collar replacement. The fasteners were installed in accordance with [3] into an SAE4340 steel tensile test fixture and tested monotonically to failure. This test fixture is illustrated in Figure 3. The ultimate tensile strength and failure mode was recorded for each specimen.

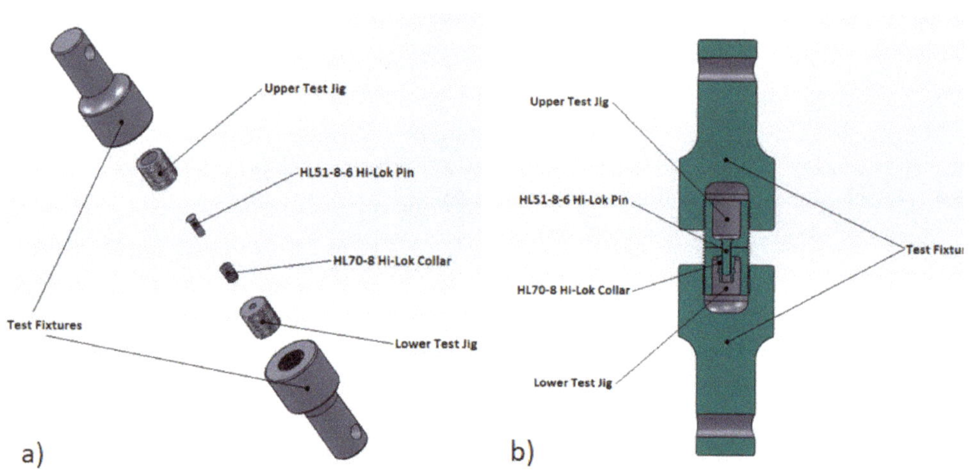

Figure 3. Static tension test assembly (**a**) exploded view and (**b**) cross section of the assembly.

2.5. Static Tension with Collar Replacement

Following completion of the baseline tests, static tensile tests were conducted on fasteners subjected to multiple collar replacements. Collars were removed and re-installed in accordance with [3] for a total of five collar replacements. The loosening and tightening torques for all collars are measured using a digital torque wrench to enable calculation of the clamping force of each collar in accordance with [11]. Following the fifth and final collar replacement, the Hi-Lok fasteners were tested monotonically to failure to determine the ultimate tensile strengths and critical failure modes for each specimen.

2.6. Baseline Shear Fatigue

Tests were conducted to calibrate load levels and determine the baseline fatigue life for the specimens with no collar replacement. The fasteners were installed in accordance with [3] into single-fastener lap joints made from 7075-T6 aluminum sheet. The assembled specimens were installed into a servo-hydraulic test system and subjected to constant-amplitude loading with a load ratio, $R = 0$ and a maximum shear fatigue load of 900 lb$_f$ (4.0 kN) based on the allowable shear load for these fasteners[4], calibrated to obtain an average fatigue life of around 500,000 cycles. Shear fatigue test specimens used the configuration shown in Figure 2a. Installed and failed specimens are shown in Figure 4.

(a)   (b)

Figure 4. Shear fatigue test specimen (**a**) loaded in test frame and (**b**) failed in fatigue.

2.7. Shear Fatigue with Collar Replacement

Following completion of the baseline tests, shear fatigue tests were conducted on specimens subjected to multiple collar replacements. Fasteners were installed in accordance with [3] into single-fastener lap joints made from 2 overlapping strips of 7075-T6 aluminum sheet. The test fixture is illustrated in Figure 4a. The assembled specimens were installed into a servo-hydraulic test system and subjected to a constant-amplitude shear fatigue load, as calibrated during the baseline fatigue tests. The Hi-Lok collars were removed and re-installed in accordance with [3] after a fixed number of load cycles. The baseline fatigue tests gave a mean fatigue life of approximately 560,000 cycles with a standard deviation of 0.159 on logarithmic fatigue life. Assuming log-normal distribution of fatigue life values, the 98% confidence level occurs at roughly 2 standard deviations below the mean life, or approximately 250,000 cycles. This was considered to be the "service life" of the joint. Hence, the experimental shear fatigue test specimens were subjected to five collar changes at intervals of 45,000 cycles, based on the lower 98 % confidence interval of the baseline fatigue life.

The loosening and tightening torques for all collars were measured using a digital torque wrench to enable calculation of the clamping force of each fastener in accordance with [11].

2.8. Statistical Analysis

In order to assess and quantify experimental differences between test specimens, simple statistical analysis was applied to the results. Welch's t-tests were performed to assess the statistical significance of any differences in static strength and fatigue life between baseline and experimental test specimens. Where multiple samples were assessed against each other, such as for the analysis of clamping force results, single-factor Analysis of Variance (ANOVA) was carried out to determine the statistical significance of any differences between sample means. In both cases, probability values were calculated to determine the statistical likelihood of having obtained the observed results through random statistical distribution within the sample.

3. Results

The experimental data collected in this study are included in tabulated form in Appendix A.

3.1. Static Shear Strength

For the static shear tests, nine baseline specimens and ten experimental specimens were tested to failure. Baseline and repeated collar replacement test results for static shear strength are compared in Figure 5a. All measured shear strengths exceed the published shear strength of 4650 lb$_f$ (20.7 kN) for the HL19-8-6 fastener pin [13]. All test specimens failed in shear of the fastener pin, as shown in Figure 5b.

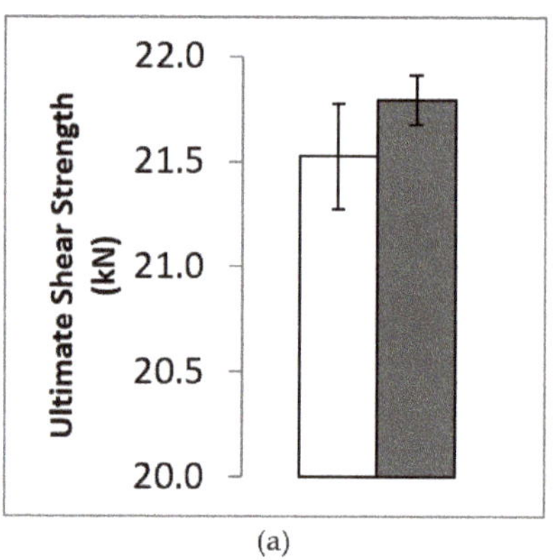
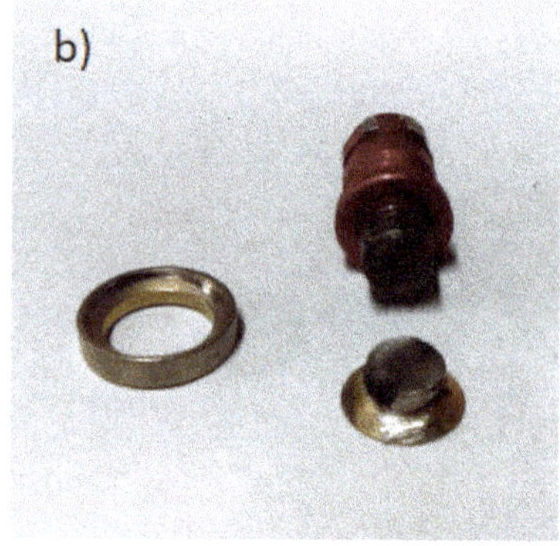

Figure 5. (**a**) average static shear strength of fastener pins: Baseline open symbol, collar replacement filled symbol (error bars represent ±1 standard deviation); and (**b**) shear failure of fastener pin (washer shown at left).

3.2. Static Tensile Strength

For the static tension tests, ten baseline specimens and ten experimental specimens were tested to failure. Baseline and repeated collar replacement results for static tensile strength are compared in Figure 6a. All measured tensile strength values exceed the published tensile strength of 3000 lb$_f$

(16.5 kN) for the HL70-8 collar [14]. All test specimens failed in tear-out of the aluminum collar, as shown in Figure 6b.

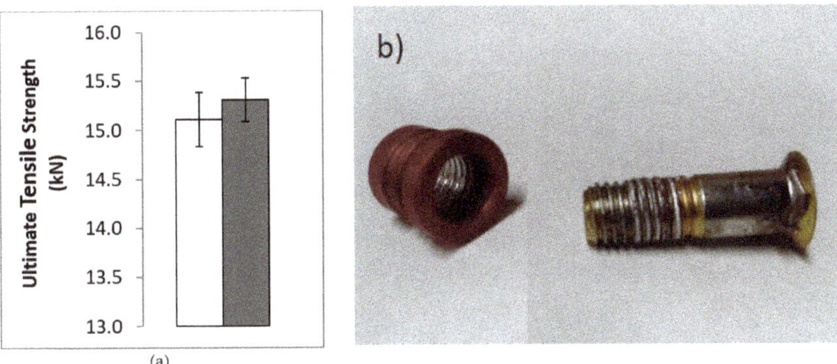

Figure 6. (**a**) Average static tensile strength of fasteners: Baseline open symbol, collar replacement filled symbol (error bars represent ±1 standard deviation); and (**b**) collar failure of fastener assembly.

3.3. Shear Fatigue Life

For the shear fatigue tests, six baseline specimens and five repeated collar replacement specimens were tested to failure in a servo-hydraulic test system. Specimens were subjected to a maximum shear fatigue load, $P_s = 900$ lb$_f$ (4.0 kN) with a load ratio, $R = 0$. Baseline and repeated collar replacement test results for shear fatigue life are compared in Figure 7. All test specimens failed in fatigue of the aluminum substrate, as shown previously in Figure 4b.

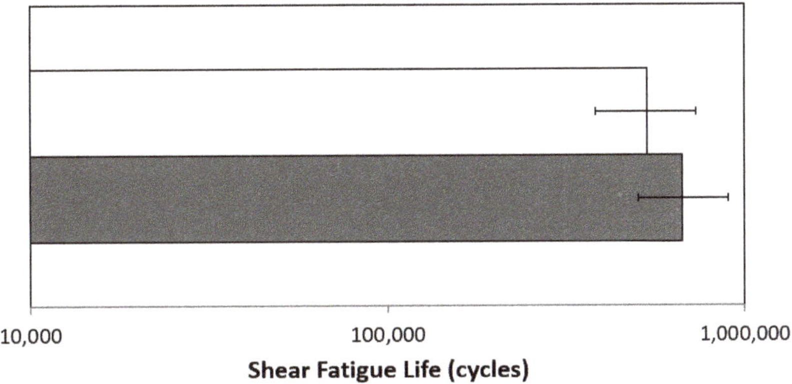

Figure 7. Average shear fatigue life of specimens: Baseline open symbol, collar replacement filled symbol (error bars represent ±1 standard deviation, assuming a log-normal fatigue life distribution).

3.4. Clamping Force

For each of the collar replacement tests in static shear, static tension, and shear fatigue, the clamping force was estimated from the measured tightening and loosening torques, as described previously.

Clamping force results for the static shear tests are summarized in Figure 8a. A time interval of approximately 30 days elapsed between the initial collar installation and removal (collar 1), and significant time-dependency was evident from the initial clamping force results. This was due to changes in the coefficient of friction between the collar and the substrate due to material relaxation. To mitigate these effects, all subsequent collar installations and removals were conducted without delay, eliminating any experimental uncertainty related to time-dependent evolution of the clamping force. Statistical analysis of the results was conducted using experimental data from collars 2 to 5 only, due to the unreliable time-dependent results from the first collar replacement.

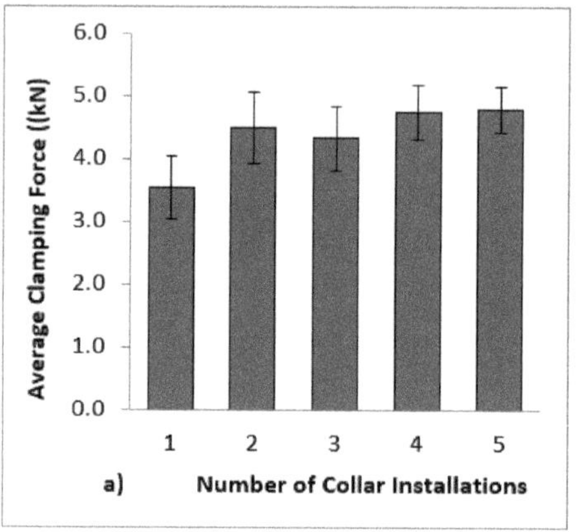

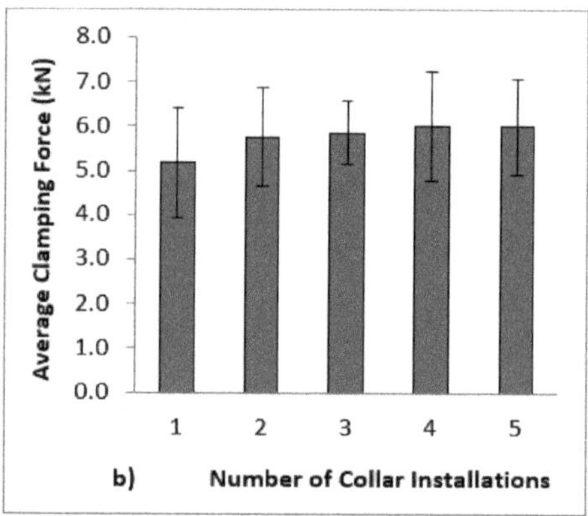

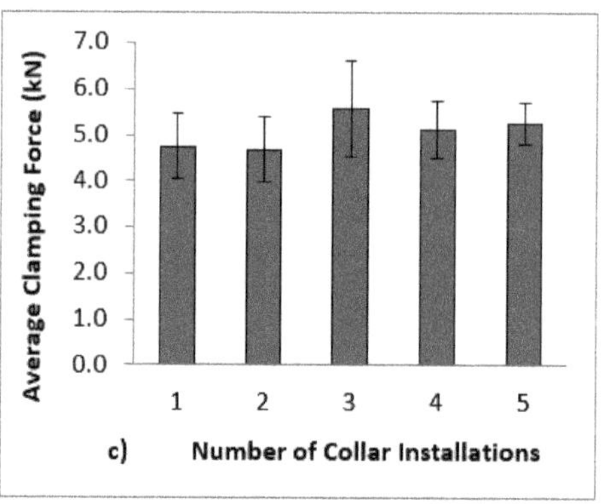

Figure 8. Average clamping force vs. number of collar installations for (**a**) static shear tests; (**b**) static tension tests and (**c**) shear fatigue tests (error bars represent ±1 standard deviation).

Clamping force results for the static tensile tests are summarized in Figure 8b. There were a small number of outliers in the initial collar replacement (collar 1) torque measurements, caused by an unfortunate rotation of the test jig within the test fixture during collar installation. For subsequent collar installations, the tensile test specimens were tightly torqued into the fixture to prevent rotation.

Note that a direct comparison between static shear (Figure 8a) and static tension (Figure 8b) results is not appropriate, given the different compliance of the steel and aluminum substrate of the shear and tensile test specimens. Clamping force results for the shear fatigue tests are summarized in Figure 8c.

4. Discussion

4.1. Static Shear Strength

All static shear specimens exceeded the minimum published single shear strength of 4650 lb_f (20.8 kN) for the HL19/51-8 fastener pin [13]. Between all specimens, the average ultimate shear strength is 4871 lb_f (21.7 kN), exceeding published values with a percent difference of 8.2%. For the specimens subjected to repetitive collar replacement, a 1.2% increase in average shear strength is observed, relative to the baseline fastener tests. While this increase is small, it is statistically significant due to the extremely narrow scatter of experimental test results, with a t-test probability value of 0.014. This value implies only a 1.4% probability of achieving this result through random statistical variation within the sample. Because there is no clear mechanism to justify an increase in static shear strength of the fastener pin as a result of collar replacement, this result is assumed to represent a statistical anomaly rather than an actual increase in static shear strength.

4.2. Static Tensile Strength

All static tensile test specimens exceed the minimum published tensile strength of 3000 lb_f (16.5 kN) for the HL70-8 collar [10]. Between all specimens, the average ultimate tensile strength is 3419 lb_f (15.2 kN), exceeding published values with a percent difference of 14.0%. For the specimens subjected to repetitive collar replacement, a 1.3% increase in average tensile strength is observed, relative to the baseline fastener tests. This increase is similar in magnitude to the increase seen for static shear strength, but is less statistically significant due to the wider scatter of the observed tensile strength results, with a t-test probability value of 0.092.

Due to the lower tensile strength of the aluminum collar, as compared to the steel fastener pins, all static tensile test specimens are critical in collar failure. Even with repetitive collar replacement, the static tensile strength of the fastener pin does not become the limiting factor for the overall tensile strength of the fastener. Because there is no clear mechanism to justify an increase in static tensile strength as a result of collar replacement, this result is not believed to represent a meaningful increase in static tensile strength. Regardless, there is no decrease in static tensile strength as a result of repetitive collar replacement for the fasteners in this test.

While static tensile tests are conducted in steel (rather than aluminum) test fixtures due to fatigue and strength requirements, this is seen to be a conservative approach. Estimated clamping force in the steel test specimens is found to be roughly 20% higher than for the aluminum shear specimens, due to the higher stiffness of the steel substrate. Mechanical damage due to collar replacement should therefore be higher in the steel test fixtures than in the simulated aluminum aircraft structure, providing a conservative estimate for the impact of collar replacement on static tensile strength.

4.3. Shear Fatigue Life

For the shear fatigue specimens subjected to repetitive collar replacement, a 1.7% increase in logarithmic fatigue life is observed relative to the baseline test specimens, representing an increase in average fatigue life from 560,000 to 682,000 cycles. Given the typically observed scatter in fatigue test results, this small experimental increase does not represent a statistically significant variation in the fatigue life of the joint, with a t-test probability value of 0.265. There is no decrease in shear fatigue life as a result of repetitive collar replacement for the fastened joints in this test.

Due to the much lower fatigue strength of the 7075-T6 aluminum substrate, as compared to the 160 ksi (1100 MPa) steel fastener pins, all shear fatigue test specimens failed by fatigue and fracture of the aluminum sheet. The shear fatigue life of the fastener pin does not become the limiting factor for the overall shear fatigue life of the specimen.

The applied 900 lb_f (4.0 kN) constant-amplitude maximum shear fatigue load for these tests is 140% of the design limit load for the fasteners at Critical Point 16 on the CP140 aircraft [10]. While test load levels are calibrated to ensure that fracture of the specimens occurred within an acceptable number of cycles, the resulting loads are not directly reflective of the actual fatigue strength or behaviour of the simulated aircraft structure.

4.4. Clamping Force

For static shear, static tension, and shear fatigue tests, no decrease in average clamping force is observed as a function collar replacement. Between all collar changes, average clamping forces for each test compare relatively well, with single-factor ANOVA probability values of 0.124, 0.482, and 0.280, for static shear, static tension, and shear fatigue tests, respectively. These results indicate no statistically significant variation in clamping force as a function of collar replacement.

In aluminum substrate, average clamping force of the fastener is found to be 1095 lb_f (4.9 kN) between static shear and shear fatigue tests. For the static tensile tests, conducted in steel fixtures, the average clamping force of 1296 lb_f (5.7 kN) is 18.4% higher, due to the higher stiffness of the steel substrate. For all tests, estimated clamping force results are highly consistent, with standard deviations well below the 17% uncertainty expected from [7].

5. Conclusions

In this project, Hi-Lok fasteners were subjected to multiple collar replacements and tested to failure under static loading and constant-amplitude fatigue loading. The aim of this research was to determine the effect of repetitive collar replacement on the residual strength and fatigue life of a retained Hi-Lok-type fastener pin. Experimental tests were conducted for static shear, static tension, and shear fatigue. In each case, the results from baseline test specimens were compared to those for identical specimens subjected to five collar replacements, simulating periodic replacement of the fastener collar in a section of fatigue-prone structure for scheduled NDI over the life of an aircraft.

From these tests, static shear results showed no decrease in the ultimate shear strength of the fastener pin as a function of collar replacement. Static tensile results showed no decrease in the ultimate tensile strength of the fastener as a function of collar replacement, with failure of the aluminum collar remaining the critical failure mode. Similarly, shear fatigue results showed no decrease in the shear fatigue life of the fastened joint as a result of collar replacement, with fracture of the aluminum substrate remaining the critical failure mode. For static shear, static tension, and shear fatigue tests, estimated clamping force was highly consistent between specimens and no decrease in clamping force was observed as a function of collar replacement.

The experimental results from this project suggest that deformations and mechanical damage resulting from a limited number of collar replacements over the life of an aircraft do not adversely affect the residual static strength, shear fatigue life, or clamping force of retained fastener pins. Although experimental results support retention of the fastener pin as a safe practice, important practical considerations such as the elevated risk of mechanical damage to the fastener pin and potential corrosion issues linked to repetitive collar removal were outside the scope of this project.

Author Contributions: D.F.H. performed the research in partial fulfillment of his MEng. degree in aerospace vehicle design with D.L.D. as the project supervisor and administrator. The individual contributions of the authors were as follows: Conceptualization, D.F.H. and D.L.D.; methodology, D.F.H.; validation, D.F.H.; formal analysis, D.F.H.; investigation, D.F.H.; resources, D.F.H. and D.L.D.; data curation, D.F.H.; writing—original draft preparation, D.F.H.; writing—review and editing, D.L.D.; supervision, D.L.D.; project administration, D.L.D.; funding acquisition, D.L.D.

Funding: Financial support of this research through the Department of National Defence, AERAC and the Natural Sciences and Engineering Research Council of Canada Discovery Grant # 239174 is gratefully acknowledged.

Conflicts of Interest: The authors declare no conflict of interest.

Nomenclature

A = dimensionless constant
D = diameter (fastener or hole)
d_2 = basic pitch diameter of the thread
d_i = inner bearing diameter of the nut
d_o = outer bearing diameter of the nut
E = elastic modulus
F = clamping force
F_{su} = ultimate shear stress
F_{tu} = ultimate tensile stress
p = pitch of the threads
P_s = shear load
P_{su} = ultimate single shear strength
P_{tu} = ultimate tensile strength
T_{on} = tightening torque
T_{off} = loosening torque
β = half-included angle of the threads
ε = strain
μ_t = coefficient of friction for the thread
μ_e = coefficient of friction for the nut face

Appendix A. Tabulated Experimental Values

The experimental data recorded for all specimens in this study are included in this Appendix A.

Table A1. Experimental data for static shear strength of fastener pins—Figure 5.

Ordered Specimens	Ultimate Shear Strength (kN)	
	Baseline Tests	**Collar Replacement Tests**
1	21.15	21.64
2	21.24	21.64
3	21.39	21.66
4	21.45	21.74
5	21.46	21.81
6	21.65	21.83
7	21.67	21.85
8	21.84	21.88
9	21.87	21.90
10	-	21.99

Table A2. Experimental data for static tensile strength—Figure 6.

Ordered Specimens	Ultimate Tensile Strength (kN)	
	Baseline Tests	**Collar Replacement Tests**
1	14.62	14.91
2	14.83	14.93
3	14.83	15.26
4	15.08	15.32
5	15.09	15.35
6	15.24	15.37
7	15.26	15.40
8	15.32	15.43
9	15.40	15.50
10	15.43	15.61

Table A3. Experimental data for shear fatigue life—Figure 7.

Ordered Specimens	Shear Fatigue Life	
	Baseline Tests	**Collar Replacement Tests**
	Cycles	Cycles
1	299,375	477,080
2	459,220	540,172
3	478,433	712,686
4	579,463	833,426
5	679,867	846,124
6	864,655	-

Table A4. Experimental data for clamping force for static shear tests—Figure 8a.

Ordered Specimen	Estimated Clamping Force (kN)				
	Collar 1	**Collar 2**	**Collar 3**	**Collar 4**	**Collar 5**
1	2.16	3.61	3.37	4.09	4.27
2	2.48	3.69	3.74	4.26	4.44
3	3.17	4.19	3.97	4.55	4.49
4	3.51	4.33	4.19	4.55	4.55
5	3.51	4.38	4.46	4.75	4.68
6	3.61	4.75	4.53	5.03	4.75
7	3.66	4.76	4.55	5.10	5.07
8	3.72	4.79	4.59	5.15	5.10
9	4.61	5.01	4.85	5.35	5.16
10	5.10	5.45	5.05	-	5.40

Table A5. Experimental data for clamping force for static tensile tests—Figure 8b.

Ordered Specimen	Estimated Clamping Force (kN)				
	Collar 1	**Collar 2**	**Collar 3**	**Collar 4**	**Collar 5**
1	4.18	3.81	5.03	4.18	4.64
2	4.36	4.90	5.18	4.88	4.79
3	4.93	5.03	5.32	5.33	4.90
4	5.03	5.23	5.48	5.48	5.67
5	5.33	5.79	5.80	5.77	5.72
6	5.70	5.95	5.95	5.80	6.39
7	6.67	6.20	5.99	6.34	6.45
8	-	6.42	6.12	6.82	6.62
9	-	6.45	6.24	6.93	6.82
10	-	7.86	7.49	8.57	8.05

Table A6. Experimental data for clamping force for shear fatigue tests—Figure 8c.

Ordered Specimen	Estimated Clamping Force (kN)				
	Collar 1	Collar 2	Collar 3	Collar 4	Collar 5
1	4.29	3.52	4.18	4.39	4.90
2	4.41	4.53	5.28	4.68	5.05
3	4.64	4.88	5.65	5.13	5.07
4	4.75	5.20	5.70	5.48	5.23
5	5.68	5.30	7.11	5.92	6.02

References

1. Lisi Aerospace. *Hi-Lok Collar*. Available online: http://www.lisi-aerospace.com/products/fasteners/internally-threaded/collars/Pages/nut-hi-lok.aspx (accessed on 5 March 2019).

2. Hi-shear Corporation. *Hi-Lok/Hi-Tigue Fastening Systems: Installation Instructions*; Hi-Shear Corporation: Torrance, CA, USA, 1991.

3. C-12-140-012/TR-001-Repair Instructions-CP140 Aurora/CP140A Arcturus-Structural Volume 1.

4. Fahr, A.; Wallace, W. *Aeronautical Applications of Non-Destructive Testing*; DEStech Publications, Inc.: Lancaster, PA, USA, 2014.

5. Jones, R.; Molent, L.; Pitt, S. Understanding Crack Growth in Fuselage Lap Joints. *Theor. Appl. Fract. Mech.* **2008**, *49*, 38–50. [CrossRef]

6. Director General of Aerospace Engineering and Project Management. *RMC 2005-007-SLA, -DGAEPM 482147 Request for Research, Scientific, and Technical Assistance from Royal Military College*; DGAEPM: Ottawa, ON, Canada, 2006.

7. Heshavanarayana, S.; Smith, B.L.; Gomez, C.; Caido, F. Fatigue-Based Severity Factors for Shear-Loaded Fastener Joints. *J. Aircr.* **2010**, *47*, 81–191.

8. Li, C.; Su, H. The Influence of Pre-tension Load in the Composite Laminate Fastener Joint During Repeated Tightening. In Proceedings of the 21st International Conference on Composite Materials, Xi'an, China, 20–25 August 2017.

9. IMP Aerospace. *DWG # 939529-CA40*; IMP Aerospace: Enfield, NS, Canada, 2010.

10. Lockheed Martin Corporation. *STP52-726*; Lockheed Martin Corporation: Bethesda, MD, USA, 2010.

11. LISI Aerospace. *HL19 Hi-Lok Pin Data Sheet*. Available online: http://www.lisi-aerospace.com/products/Pages/Fasteners-Catalog.aspx (accessed on 5 March 2019).

12. Eccles, W. *A New Approach to the Checking of the Tightness of Bolted Connections*; Lubrication, Maintenance, and Tribotechnology; Paper Number: L144056; LUBMAT 2014: Manchester, UK, 2014.

13. Eccles, W. *A New Approach to the Tightness Checking of Bolts*. Fastener + Fixing Magazine; Issue 90. 2014. Available online: http://boltscience.com/pages/a-new-approach-to-the-tightness-checking-of-bolts.pdf (accessed on 5 March 2019).

14. LISI Aerospace. *HL70 Hi-Lok Collar Data Sheet*. Available online: http://www.lisi-aerospace.com/products/Pages/Fasteners-Catalog.aspx (accessed on 5 March 2019).

Article

Fatigue Life Prediction of Steam Generator Tubes by Tube Specimens with Circular Holes

Qiwei Wang [1], Junfeng Chen [1], Xiao Chen [1], Zengliang Gao [1,2] and Yuebing Li [1,*]

[1]   Institute of Process Equipment and Control Engineering, Zhejiang University of Technology, Hangzhou 310032, China; wangqiweide@foxmail.com (Q.W.); jfchappy@163.com (J.C.); iversonchenxiao@163.com (X.C.); zlgao@zjut.edu.cn (Z.G.)

[2]   Engineering Research Center of Process Equipment and Remanufacturing, Ministry of Education, Hangzhou 310032, China

*   Correspondence: ybli@zjut.edu.cn; Tel.: +86-15869128380

Received: 30 January 2019; Accepted: 10 March 2019; Published: 12 March 2019

Abstract: Heat exchangers manufactured from Inconel 690 tubes are widely used for steam generators in nuclear power plants. Inconel 690 tubes have suffered failures of fatigue fracture due to flow induced vibration. It is difficult to obtain the fatigue life of the tube directly since the conventional fatigue test would potentially cause end fatigue failure due to the stress concentration at the clamp end. In this study, a thin-walled Inconel 690 tube with circular hole is designed to deduce the fatigue life of smooth tube based on the notch fatigue life prediction technology. Firstly, the local stress and strain distributions around the hole based on the finite element analysis are discussed. Local stress-strain is calculated and compared with Neuber's ruler. Meanwhile, fatigue life tests using tube specimens with circular holes are carried out. Finally, based on the best-fitted fatigue life curve of Inconel 690 alloy, the fatigue life of tube specimen is estimated from the local strain according to Neuber's ruler. The results show that the local stress and strain estimated by Neuber's ruler are basically consistent with those obtained by finite element analysis. Compared with the average fatigue life of nickel-based alloy, the new predicted equivalent fatigue life of heat Inconel 690 transfer tube with a hole is higher. The Inconel 690 heat transfer tube has better fatigue performance.

Keywords: tube specimen with hole; fatigue life; local strain; Inconel 690 tube

1. Introduction

Inconel 690 heat transfer tubes, with excellent heat transfer performance, have been widely used in nuclear steam generators, ultra-supercritical boilers, and other important equipment in nuclear power plants. Under the action of fluid excitation, these heat exchange devices are prone to cause vibration of the heat transfer tube, resulting in the fatigue fracture failure of heat transfer tubes [1,2]. Moreover, the heat transfer tube is subject to many degradation mechanisms, e.g., corrosion, wear, and fatigue. A large number of studies focused on the corrosion behaviour of tube materials with service conditions, i.e., high temperature and high pressure water. It should be noted that fatigue failure of tubes must be considered during the design and operation of steam generators. Fatigue design curve is necessary to conduct the fatigue design of tubes, which is usually derived from large tests of similar materials under atmospheric environment. To consider the effect of corrosion degeneration, a correction factor is usually adopted to revise the fatigue life under atmospheric [3,4]. Therefore, it is necessary to perform fatigue tests for tubes.

The Inconel 690 heat transfer tube for nuclear steam generators is a uniform thin-walled tube structure with a diameter of Ø17.48 mm and a wall thickness of only 1 mm. If the fatigue test is performed directly with the actual thinned wall tube, the stress concentration will easily occur in the clamping ends leading to fatigue failure [5]. Thus, the obtained fatigue life is not the actual fatigue life of the heat transfer tube. To resolve this issue, we tried several tube specimens to carry out fatigue tests. In this work, an attempt by tube specimens with holes was made to deduce fatigue life for tube based on the notch fatigue life prediction technology.

Some prediction models for the notch effect on fatigue live had been proposed to evaluate the fatigue life of components with notches. Generally, an effective parameter is introduced to link the fatigue lives between the notch and smooth specimens, such as the local stress-strain method [6–10], the critical distance method [11,12], and stress field strength method [13–15]. One of the most popular methods to analyse notch fatigue problems was formulated by Neuber, who stated that the geometric mean value of both the stress and strain concentration factors is constant at any load state, and equals the elastic stress concentration factor [16]. This method has not only been successfully applied to a wide range of engineering problems (e.g., [17–19]) but among the fatigue life assessment methods recommended by the International Institute of Welding [9].

In this study, a notched specimen of Inconel 690 heat transfer tube was designed and tested under cyclic loading. Based on the finite element analysis, the local stress and strain distributions around the notch are obtained and compared with Neuber's ruler. The fatigue life curve of Inconel 690 material is used to estimate the fatigue life of tubular specimens with hole based on local strain according to Neuber's ruler. The fatigue life is verified by the test results. In our experiments, the fatigue life prediction method for heat transfer tubes provides a basis for engineering design assessment.

2. Experimental Procedure and Results

2.1. Experimental Procedure

The chemical composition of Inconel 690 alloy thin-walled tube commonly used in nuclear power steam generators is shown in Table 1, and the mechanical properties are given in Table 2.

Table 1. Chemical compositions of Inconel 690 (wt%).

Ni	Cr	Fe	C	Mn	Si	Cu	S
58.0 min.	27.0–31.0	7.0–11.0	0.05 max.	0.50 max.	0.50 max.	0.5 max.	0.015 max.

Table 2. Tensile properties of Inconel 690 tube.

Item	Young's Modulus (MPa)	Yield Strength (MPa)	Tensile Strength (MPa)
Experimental results	2.12×105	286	702
Standard requirement	-	240	586

The specimen was directly taken from the straight tube section of the steam generator tube with the outer diameter of 17.48 mm and a wall thickness of only 1.01 mm. The detailed specifications of the sample are shown in Figure 1. The length of the specimen is 150 mm, a through hole with different diameters (d = 2 mm and 4 mm) in the middle of tube were made, as shown in Figure 1.

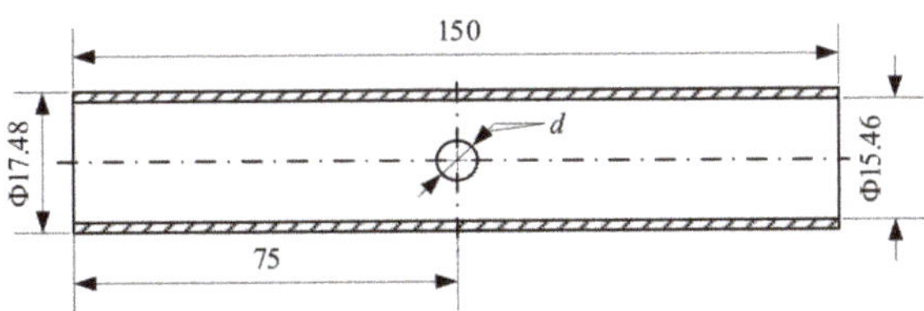

Figure 1. Test specimen for thin wall tube with holes. Unit: mm.

The tests were carried out on the Intsron 8802 hydraulic servo fatigue testing machine (Illinois Tool Works Inc., Norwood, MA, USA) under the stress control mode and strain control mode respectively.

The test rig is shown in Figure 2a. To avoid the deformation due to the clamping, a plug is designed and inserted the tube, as shown in Figure 2b. In both stress control and strain control, symmetrical loading is adopted, that is, the stress ratio is −1.

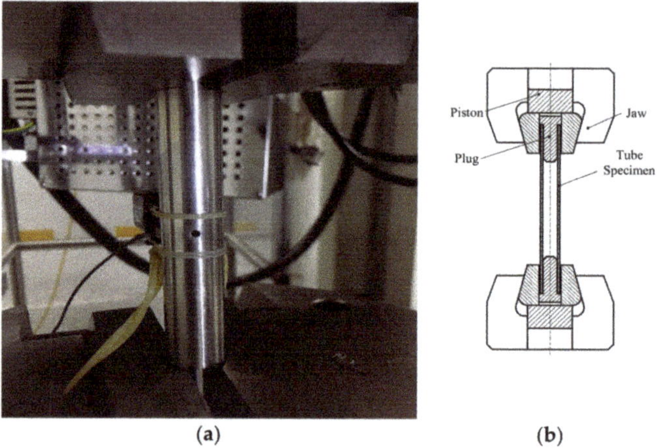

Figure 2. Picture of the test device: (**a**) Test rig, and (**b**) diagrammatic sketch of clamping technique.

2.2. Results and Analysis

The test results for the tube specimens with holes under different cyclic loadings are shown in Table 3. For the strain control test, the fatigue life is the number of cycles when the load decreases by 5%, while the fatigue life for the stress control test is the number of cycles corresponding to a 5% increase in displacement.

Table 3. Fatigue lives of tubes with holes under different loading with stress/strain control mode.

Diameter of Holes	Control Method	Load Level	Test Life
4 mm	Strain control	±0.2%	478, 886
2 mm	Stress control	±250 MPa	11,722, 13,300
2 mm	Stress control	±200 MPa	44,200, 58,800

For a hole in tubular specimens, the fatigue cracks initiate on the hole edge and extend perpendicular to the axial direction. In the thickness direction, the crack surface exhibits different angles with the crack growth, as shown in Figure 3. Figure 3a gives a macroscopic view of the crack which is perpendicular to the loading direction. Figure 3b shows the fracture appearance near the hole and a diagrammatic sketch is drawn in Figure 3c. It can be observed that the crack initiates on the plane of maximum shear stress and is at an angle to the diameter of the hole. The crack initiation is on the plane of maximum shear stress and is at a certain angle to the radial direction of the hole. As the crack propagates, the effect of the hole is weaker and the crack tip tends to be in a planar stress state. Therefore, in the region far from the hole, a sharp oblique fracture is exhibited. Thus, the fracture can be divided into three regions. The first region, from the edge of the notch to the blue line, corresponds to crack formation on maximum shear plane where slip occurs. The fracture surface in the first region presents a plane with normal at 45° to the specimen axial. As the crack propagates, the effect of the notch is weaker and the stress state is changed. The normal of crack face is switching with the stress state, as shown in the section region between the blue line and green line. When the crack continues propagating, a third region is displayed with slant appearance. In this region, notch effect can be ignored, which is similar with smooth specimen. In addition, a microscopic morphology appears fatigue striations, as shown in Figure 3d.

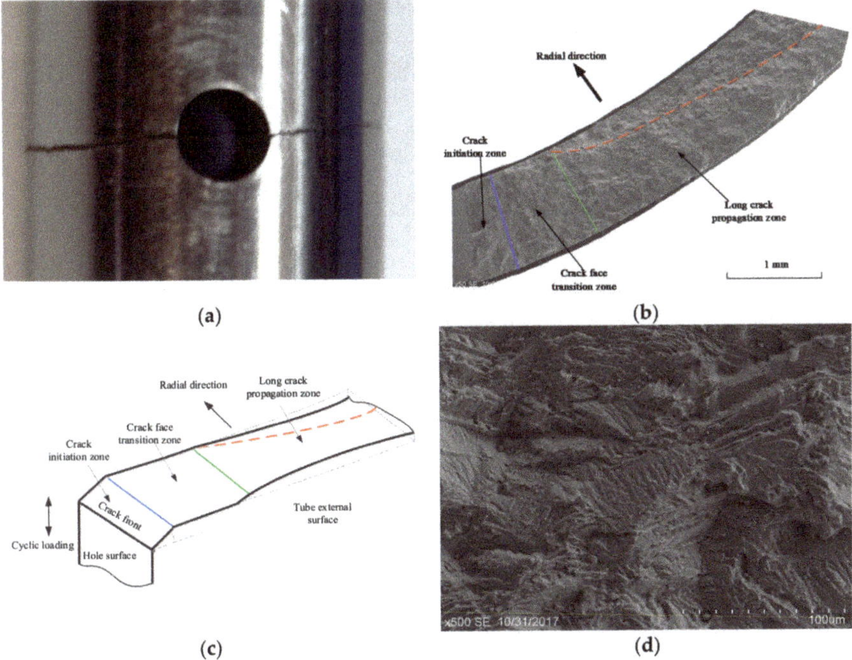

Figure 3. Fracture appearance for tube with holes: (**a**) Macroscopic view of the crack, (**b**) fracture appearance, (**c**) diagrammatic sketch, and (**d**) microscopic morphology.

3. Evaluation of Stress and Strain near the Hole

3.1. Finite Element Model

According to the symmetry, a 1/8 model of the notched specimen was constructed using finite element software ABAQUS (Dassault Systèmes, version 6.12, Vélizy-Villacoublay Cedex, France). 20-node hexahedral element C3D20R was selected. Consider the stress concentration at the root of the notch, local refinement of the area near the notch was conducted, as shown in Figure 4. The element size along the notch is about 0.04 mm × 0.05 mm × 0.05 mm. Symmetric boundaries are applied to the three symmetric planes of the model, and pressure load is applied to one end of the tube.

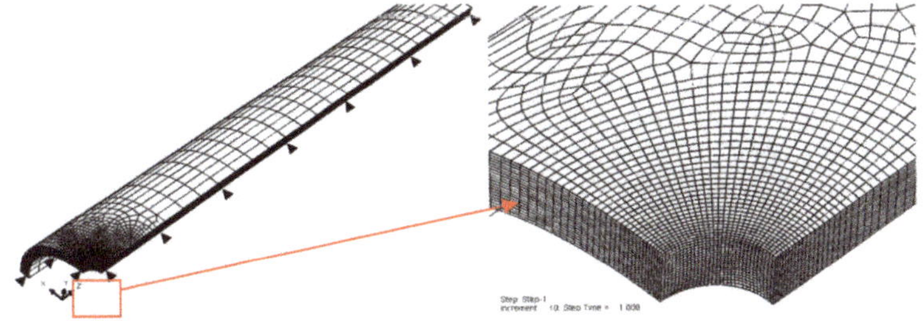

Figure 4. Finite element model.

In order to obtain the local stress-strain state around the notch, elastic analysis and elastoplastic analysis were carried out respectively. Meanwhile, the cyclic stress-strain relation of material for elastoplastic analysis is described by Ramberg-Osgood model [20]:

$$\frac{\Delta\varepsilon}{2} = \frac{\Delta\sigma}{2E} + \left(\frac{\Delta\sigma}{2K'}\right)^{1/n'} \tag{1}$$

where the cyclic strength coefficient $K' = 424.92$ MPa, the cyclic strain-hardening exponent $n' = 0.129$.

3.2. Elastic Analysis

The elastic stress concentration factor is one of the important parameters in the local stress-strain method to estimate the fatigue life of notched specimens. Therefore, the elastic finite element analysis of the thin-walled tube with hole is carried out. To investigate the local stress-strain, three paths are defined as shown in Figure 5. Path P1 locates at the external surface along the circumference of the hole, while P2 locates at the hole edge along the wall and P3 on the mid-thickness wall along the radial path of the hole. The stress distributions on different paths are obtained. In the analysis, the stress concentration factor, which is the ratio of stress level to applied stress, is used to describe the stress field.

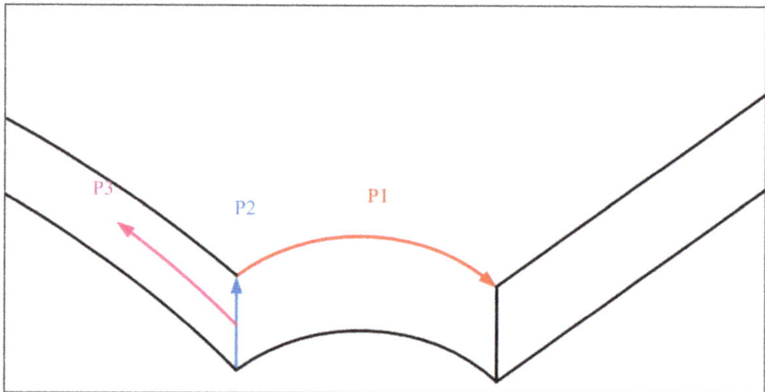

Figure 5. Path diagrammatic sketch.

The equivalent stress and maximum principal stress distribution along the outer circumference on the tube wall (P1 path in Figure 5) are shown in Figure 6. As shown in Figure 6, the stress concentration is perpendicular to the load direction, i.e., the 0° position in the figure. At this position, the stress distribution in the wall thickness direction (P2 path in Figure 5) is shown in Figure 7. Due to the constraint effect [21], the maximum stress appears in the middle of the thickness, which is more than about 7% above the surface stress. At this stage, the fatigue damage in the middle portion of the wall thickness is the highest. From the fracture results of the specimen, it can be found that the crack initiation source of most of the specimens is in the middle of the thickness of the hole edge.

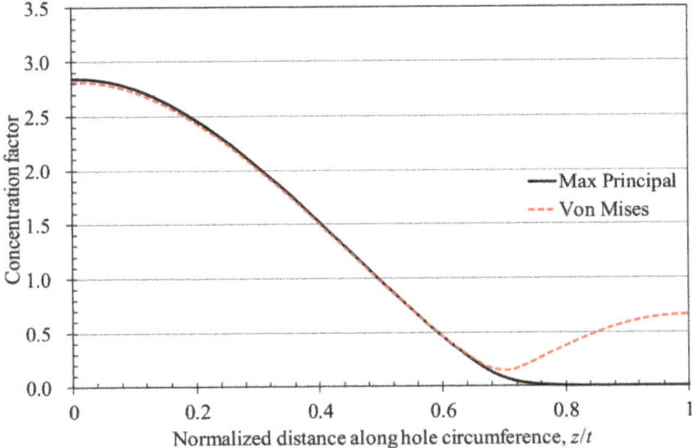

Figure 6. Distribution of Mises stress and max principal stress around the hole circumference (path P1).

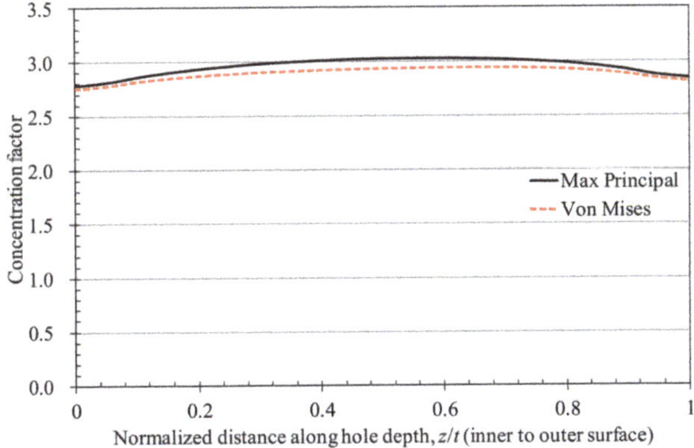

Figure 7. Distribution of Mises stress and max principal stress along the hole depth (path P2).

Figure 8 shows the stress distribution along the radial path of the hole (P3 path in Figure 3). In the figure, r is the radius of the hole. The stress decreases with the radial distance from the hole edge. However, the stress tends to be stable beyond a certain length. The difference between the Mises stress and the maximum principal stress shows a trend of increasing initially then decreasing. The greater the difference between the two, the more obvious the stress triaxiality. When the two are equal, the structure is in a uniaxial stress state. For a thin-walled tube, if the stress in the thickness direction is ignored, it would be in a plane stress state.

In the cylindrical coordinate system, the stress state at a point of the tube can be simplified as two-dimensions plane stress with axial stress, hoop stress and shear stress. Under axial loading, the maximum principal stress closely approximates the axial stress, which usually independents with the tube thickness, as shown in Figure 9. However, the hoop stress presents much more correlatively with thickness than the two other component stresses. Further, the directions of maximum shear stress change with the hoop stress component. The values of maximum shear stress near the inner surface are higher than the external surface. This phenomenon results in a more serious fatigue damage to the region of inner surface, initiating a crack easily. For the region away from notch, the stress state tends to be uniaxial. Therefore, there is a transition region for the maximum shear stress. This is consistent with the formation mechanism of the three-region-fracture from the stress state of the hole.

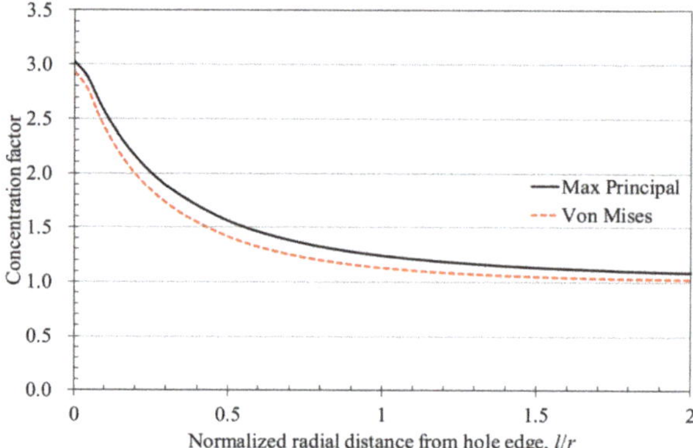

Figure 8. Distribution of Mises stress and max principal stress radically away from hole edge on specimen outer surface (path P3).

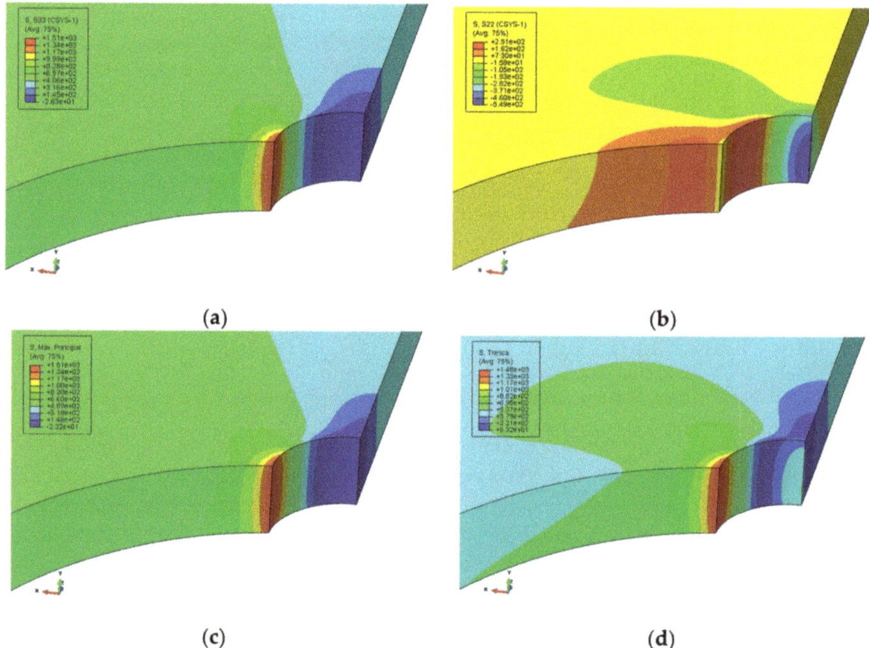

(a)   **(b)**

(c)   **(d)**

Figure 9. Stress contours from elastic finite element analysis: (**a**) Axial stress, (**b**) hoop stress, (**c**) maximum principal stress, and (**d**) maximum shear stress.

3.3. Elastoplastic Analysis

In the notched fatigue analysis, the precise local stress-strain field is the key to predicting the fatigue life of the notch. For materials, stress concentration is likely to occur near the notch which may exceed the yield strength of the material. Therefore, the elastoplastic analysis can describe local stress-strain field more accurately.

In the analysis, the Neuber's formula is also used to calculate the local stress-strain of the notch and predict the fatigue life of the notched component. The Neuber's formula [22,23] can be expressed as:

$$\varepsilon_q \sigma_q = \frac{K_{tq}^2 S^2}{E} \cdot \frac{Ee^*}{S^*} \tag{2}$$

where σ_q and ε_q are local equivalent stresses and strains, K_{tq} is the equivalent elastic stress concentration factor, S^* and e^* are plastically corrected nominal stresses and nominal strains, E is the elastic modulus, and S is the elastic nominal stress. The asterisked nominal stress and strain are corrected for nominally inelastic loadings, which are not considered in this work as the nominally applied stress is lower than the yield strength. Figure 10 compares the local equivalent strain as functions of nominal stress estimated with elastoplastic finite element analysis and Neuber's formula. As shown in Figure 10, the estimation is basically consistent with the finite element calculation result with an error less than 6%. In this work, the Neuber's method is also used to analyse the local stress-strain of the notch to estimate the fatigue life.

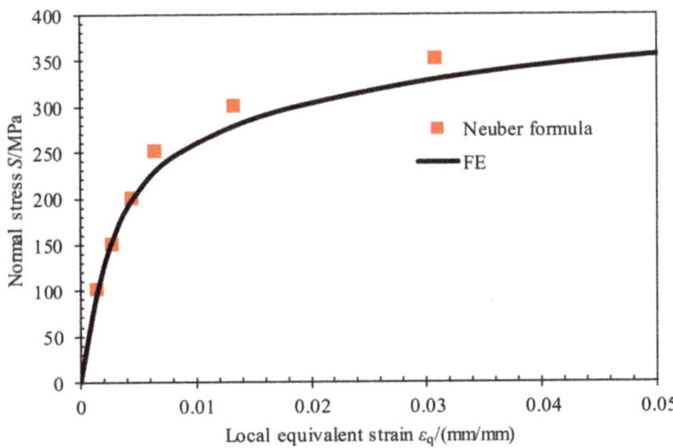

Figure 10. Neuber's rule estimations compared to finite element results for nominal stress vs. local equivalent strain.

4. Fatigue Life Analysis

Based on the cyclic stress and strain curve of materials, local equivalent strain of structure is calculated by elastoplastic finite element analysis and estimation method. The fatigue life of structure is estimated by strain fatigue life curve. Based on the results of finite element analysis, Neuber's method can be used to predict the local strain in the linear elastic range. Therefore, Neuber's formula is used to estimate the equivalent strain of the notch structure. In order to interpret notch effect on fatigue life, the elastic stress concentration coefficient is revised to the fatigue notch coefficient K_f:

$$K_f = 1 + \frac{K_t - 1}{1 + \sqrt{\rho/r}} \tag{3}$$

where K_t is the elastic stress concentration factor, r is the notch radius, and ρ is the material characteristic length. The material characteristic length can be estimated on the ultimate tensile strength σ_u of alloy by the following empirical relation [24,25]:

$$\log \rho = -\frac{\sigma_u - 134}{586} \tag{4}$$

According to the tension properties of the tube, one can obtain $\rho = 0.11$ mm. The elastic stress concentration factor is obtained by elastic finite element analysis as described in Section 3.2, where

$K_t = 2.94$ for $d = 2$ mm and $K_t = 3.54$ for $d = 4$ mm. Combining Equations (1) and (2), the local equivalent strain near the hole can be obtained by replacing equivalent elastic stress concentration factor K_{tq} with fatigue notch coefficient K_f. Then, the equivalent strain amplitude ε_a of the smooth tube can be calculated.

Figure 11 shows the equivalent strain amplitude for the test samples. Moreover, the strain fatigue life curve of nickel-base alloy materials is also plotted. The strain fatigue life curve is described with Langer equation [26]:

$$\varepsilon_a = 14.967 N_f^{-0.4053} + 0.0805 \tag{5}$$

which is usually used in the codes and standards for fatigue assessment of components containing pressure [27]. As shown in Figure 11, the equivalent fatigue life based on the notched specimen is consistent with the material fatigue life trend. Further, the equivalent fatigue life based on the notched specimen is mostly higher than the fatigue life of the material. Figure 12 shows the difference between the life prediction values and experimental values based on the material life curve. At higher load levels, these are more consistent, that is, the low-cycle life zone based on Neuber's method can predict the fatigue life of the heat transfer tube more accurately. However, as the fatigue life increases, the deviation between them increases. Moreover, the test value may fall outside the dispersion band of the double predicted life.

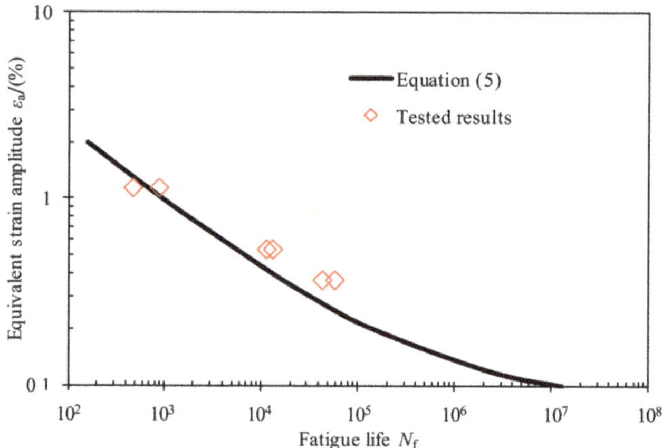

Figure 11. Fatigue life curve of tube with holes and material.

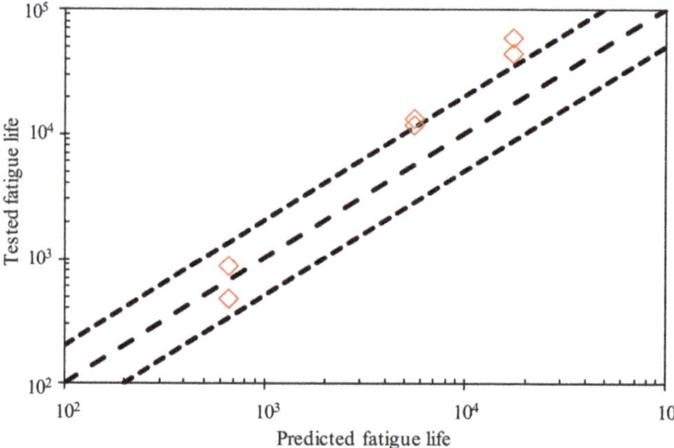

Figure 12. Comparison test result with predicted result of fatigue life of tube with holes.

This difference may be attributed to two reasons: fatigue life prediction model itself and material fatigue life curve. (1) When the prediction method is concerned, the fatigue life prediction model based on local stress-strain is expected to conservatively predict the low-cycle fatigue behaviour of the structure. However, there are still limitation in the prediction of high-cycle fatigue. (2) The fatigue life curve of the materials used in the paper is derived from the fatigue life test results of many nickel-based alloys. It can represent the mean fatigue life of nickel-based alloy materials which may differ from the smooth tube fatigue properties referred. Actually, the predicted result of this method is conservative, i.e., the fatigue life of the smooth tube involved in this paper should be higher than the average fatigue life of the nickel-based alloy material. This also indicates that the heat transfer tubes mentioned in this work should have better fatigue properties.

5. Conclusions

Aiming at the fatigue life of Inconel 690 heat transfer tube, a thin-walled tube specimen was used and tested for fatigue life. The fatigue life of heat transfer tube was predicted and analysed based on the local stress-strain field around the hole. The conclusions are as follows:

(1) Under the axial cyclic loading, the fatigue crack initiates on the boundary of the hole and propagates perpendicular to the axial of the tube. The fracture of the sample can be divided into three areas: the crack initiation zone, the fracture zone transition zone, and the long crack extension zone. In the thickness direction, the crack face exhibits different angles as the crack propagates.

(2) The elastic stress concentration factor of the tube specimen with a hole is obtained by elastic finite element analysis. Also, the stress distribution is discussed in this paper. The fatigue damage in the middle of the wall thickness is the highest, which is consistent with the test fracture analysis.

(3) The local stress-strain field of the tube specimen with a hole is obtained by elastoplastic finite element analysis. Compared with the Neuber's formula estimation results, the error between them is less than 6%. The Neuber's method can be used to estimate the local stress-strain of the notch.

(4) Based on the local stress-strain field of the tube specimen with a hole, the equivalent fatigue life of the smooth heat transfer tube was obtained and compared with the fatigue life of the raw material. Compared with the mean fatigue life of nickel-based alloy materials, the equivalent fatigue life of smooth heat transfer tubes is higher. Further, the heat transfer tubes involved should have a better fatigue performance.

Author Contributions: Conceptualization: Y.L.; formal analysis: J.C.; funding acquisition: Y.L.; investigation: Q.W., J.C., and X.C.; methodology: Z.G.; supervision: Z.G.; writing—original draft: Q.W.; writing—review and editing: Y.L.

Funding: This research was funded by National Key R&D Program of China, grant number 2018YFC0808800, and National Natural Science Foundation of China, grant number 51605435.

Conflicts of Interest: The authors declare no conflict of interest.

References

1. Auvinen, A.; Jokiniemi, J.K.; Laehde, A.; Lähde, A.; Routamo, T.; Lundström, P.; Tuomisto, H.; Dienstbier, J.; Güntay, S.; Suckow, D.; et al. Steam generator tube rupture (SGTR) scenarios. *Nucl. Eng. Des.* **2005**, *235*, 457–472. [CrossRef]
2. Diercks, D.R.; Shack, W.J.; Muscara, J. Overview of steam generator tube degradation and integrity issues. *Nucl. Eng. Des.* **1999**, *194*, 19–30. [CrossRef]
3. Japan Nuclear Energy Safety Organization Nuclear Energy System Safety Division. *Environmental Fatigue Evaluation Method for Nuclear Power Plants, JNES-SS-1005*; Japan Nuclear Energy Safety Organization: Tokyo, Japan, 2011.

4. Higuchi, M.; Sakaguchi, K.; Hirano, A.; Nomura, Y. Revised and new proposal of environmental fatigue life correction factor (Fen) for carbon and low-alloy steels and nickel base alloys in LWR water environments. In Proceedings of the PVP2006-ICPVT-11, ASME Pressure Vessels and Piping Division Conference, Vancoucer, BC, Canada, 23–27 July 2006.

5. Arora, P.; Gupta, S.K.; Bhasin, V.; Singh, R.K.; Sivaprasad, S.; Tarafder, S. Testing and assessment of fatigue life prediction models for Indian PHWRs piping material under multi-axial load cycling. *Int. J. Fatigue* **2016**, *85*, 98–113. [CrossRef]

6. Neuber, H. Theory of stress concentration for shear strained prismatical bodies with arbitrary non-linear stress-strain law. *J. Appl. Mech.* **1961**, *28*, 544–550. [CrossRef]

7. Zhang, C.C.; Yao, W.X. Typical fatigue life analysis approaches for notched components. *J. Aerosp. Power* **2013**, *28*, 1223–1230.

8. Ellyin, F.; Kujawski, D. Generalization of notch analysis and its extension to cyclic loading. *Eng. Fract. Mech.* **1989**, *32*, 819–826. [CrossRef]

9. Karakaş, Ö. Application of Neuber's effective stress method for the evaluation of the fatigue behaviour of magnesium welds. *Int. J. Fatigue* **2017**, *101*, 115–126. [CrossRef]

10. Bentachfine, S.; Pluvinage, G.; Gilgert, J.; Azari, Z.; Bouami, D. Notch effect in low cycle fatigue. *Int. J. Fatigue* **1999**, *21*, 421–430. [CrossRef]

11. Taylor, D. Geometrical effects in fatigue: A unifying theoretical model. *Int. J. Fatigue* **1999**, *21*, 413–420. [CrossRef]

12. Sun, D.; Hu, Z.D. Research of the size factor of fatigue strength base on TCD theory. *Chin. Quart. Mech.* **2015**, *36*, 288–295.

13. Adib-Ramezani, H.; Jeong, J. Advanced volumetric method for fatigue life prediction using stress gradient effects at notch roots. *Comp. Mater. Sci.* **2007**, *39*, 649–663. [CrossRef]

14. Yao, W.X. Stress field intensity approach for predicting fatigue life. *Int. J. Fatigue* **1993**, *15*, 243–246.

15. Shen, J.B.; Tang, D.L. Predicting method for fatigue life with stress gradient. *Chin. Mech. Eng.* **2017**, *28*, 40–44.

16. Branco, R.; Costa, J.; Berto, F.; Antunes, F.V. Fatigue life assessment of notched round bars under multiaxial loading based on the total strain energy density approach. *Thoret. Appl. Fract. Mech.* **2018**, *97*, 340–348. [CrossRef]

17. Sonsino, C.M.; Bruder, T.; Baumgartner, J. S-N Lines for welded thin joints—Suggested slopes and FAT values for applying the notch stress concept with various reference radii. *Weld World* **2010**, *54*, 375–392. [CrossRef]

18. Sonsino, C.M.; Fricke, W.; De Bruyne, F.; Hoppe, A.; Ahmadi, A.; Zhang, G. Notch stress concepts for the fatigue assessment of welded joints—Background and applications. *Int. J. Fatigue* **2012**, *34*, 2–16. [CrossRef]

19. Bertini, L.; Frendo, F.; Marulo, G. Fatigue life assessment of welded joints by two local stress approaches: The notch stress approach and the peak stress method. *Int. J. Fatigue* **2018**, *110*, 246–253. [CrossRef]

20. He, X.; Chen, J.; Tian, W.; Li, Y.; Jin, W. Low cycle fatigue behavior of steam generator tubes under axial loading. *Materials* **2018**, *11*, 1944. [CrossRef]

21. Guo, W. Three-dimensional analyses of plastic constraint for through-thickness cracked bodies. *Eng. Fract. Mech.* **1999**, *62*, 383–407. [CrossRef]

22. Seeger, T.; Heuler, P. Generalized application of Neuber's rule. *J. Test. Eval.* **1980**, *8*, 199–204.

23. Gates, N.; Fatemi, A. Notched fatigue behavior and stress analysis under multiaxial states of stress. *Int. J. Fatigue* **2014**, *67*, 2–14. [CrossRef]

24. Kuhn, P.; Hardrath, H.F. *An Engineering Method for Estimating Notch-Size Effect in Fatigue Tests on Steel*; NACA Technical Note; NACA: Langley Field, VA, USA, 1952.

25. Gates, N.; Fatemi, A. Notch deformation and stress gradient effects in multiaxial fatigue. *Theor. Appl. Fract. Mech.* **2016**, *84*, 3–25. [CrossRef]

26. Langer, B.F. Design of pressure vessels for low-cycle fatigue. *Trans. ASME J. Basic Eng.* **1962**, *84*, 389–399. [CrossRef]

27. Chopra, O.; Stevens, G. *Effect of LWR Coolant Environments on the Fatigue Life of Reactor Materials*; NUREG/CR-6909, ANL-12/60; U.S. Nuclear Regulatory Commission: Washington, DC, USA, 2014.

Article

Retardation of Fatigue Crack Growth in Rotating Bending Specimens with Semi-Elliptical Cracks

Martin Leitner [1],*, David Simunek [1], Jürgen Maierhofer [2], Hans-Peter Gänser [2] and Reinhard Pippan [3]

[1] Department Product Engineering, Montanuniversität Leoben, 8700 Leoben, Austria; david.simunek@unileoben.ac.at
[2] Materials Center Leoben Forschung GmbH, 8700 Leoben, Austria; juergen.maierhofer@mcl.at (J.M.); hp.gaenser@mcl.at (H.-P.G.)
[3] Erich Schmid Institute of Materials Science, 8700 Leoben, Austria; reinhard.pippan@oeaw.ac.at
* Correspondence: martin.leitner@unileoben.ac.at; Tel.: +43-3842-402-1463

Received: 15 December 2018; Accepted: 26 January 2019; Published: 1 February 2019

Abstract: This work investigates overload-induced retardation effects for semi-elliptically cracked steel round bars. The specimen geometry equals the shaft area of a 1:3 down-scaled railway axle and the material is extracted from railway axle blanks made of EA4T steel. Rotating bending tests under constant amplitude loading as well as overload tests considering overload ratios of R_{OL} = 2.0 and R_{OL} = 2.5 are conducted. The experimental results are compared to a crack growth assessment based on a modified NASGRO equation as well as the retardation model by Willenborg, Gallagher, and Hughes. The evaluated delay cycle number due to the overload by the experiments and the model shows a sound agreement validating the applicability of the presented approach.

Keywords: fatigue crack growth; overload; retardation; semi-elliptical crack; rotating bending

---

1. Introduction

In order to assess the fatigue life under variable amplitude load (VAL) scenarios, the use of the linear damage accumulation rule according to Palmgren [1] and Miner [2] still acts as a standardized approach, due to the comparably engineering-feasible applicability. However, especially the impact of load sequences, such as retardation effects by intermediate overloads, majorly affect the fatigue life under VAL, which is not properly considered applying this rule [3]. In general, the retardation effect is a physically complex phenomenon, which is influenced by several interacting factors such as the loading condition, metallurgical properties, environment, and others [4].

According to Carlson et al. [5], the most significant mechanisms affecting the retardation behavior after a single overload are based on residual stresses [6], crack deflection [7], crack closure [8], strain hardening [9] as well as plastic crack tip blunting/resharpening [10]. Several models to assess the retardation effect during VAL are available, whereas in [11], a separation in yield zone and crack closure concepts is presented. Herein, the models by Wheeler [12], Willenborg et al. [13], Porter [14], Gray and Gallagher [15], Gallagher and Hughes [16], Johnson [17], as well as by Chang et al. [18] are highlighted in the case of the yield zone concepts. For the crack closure approaches, the models by Elber [8], Bell and Creager [19], Newman [20], Dill and Staff [21], Kanninen et al. [22], Budiansky and Hutchinson [23], as well as by de Koning [24] are referred. Due to their simplicity and the advantage that model parameters can be practicably evaluated by experiments, the yield zone models are commonly applied within crack growth assessments considering retardation effects at VAL [25]. Further details regarding the crack growth behavior under VAL, as well as the application of assessment models are provided in [26].

This paper focuses on the retardation effect in case of round bars containing semi-elliptical cracks. The investigations are part of the research project Eisenbahnfahrwerke 3 (EBFW 3) [27], which holistically analyzes the residual fatigue life of railway axles in order to properly assess inspection intervals [28]. As maintained in [29], the material parameters, loading conditions, as well as the initial crack geometry act as fundamental factors influencing a safe life fatigue assessment of railway axles [30]. A study in [31] demonstrates that measured load spectra of railway axles majorly affect the residual life; hence, sequence effects such as crack retardation need to be considered within the crack growth estimation. In addition, within the EBFW 3 project, the transferability of fracture mechanical parameters from small-scale, single edged tension or bending (SET or SEB) specimens with straight crack fronts to full-scale railway axles, containing semi-elliptical cracks is investigated, see [27]. In between these two completely different geometries and dimension, a down-scaled round bar specimen, denoted as 1:3 specimen, is used to investigate the issue of transferability in detail. Further details regarding the manufacturing and testing procedure of the 1:3 scaled specimens and experimental results covering constant amplitude and overloads are provided in [32]. In summary, this paper scientifically contributes to the following research topics:

- Crack growth tests with 1:3 round bar specimens, incorporating constant amplitude loading, as well as the effect of overloads.
- Application of a yield zone model based on small-scale SEB specimen test data to assess the influence of overloads on the fatigue crack retardation of the 1:3 specimens.
- Comparison of results by experiments, modelling and evaluation of the transferability of test data from small-scale SEB specimens, with straight crack fronts to round bars containing semi-elliptical cracks.

2. Materials and Methods

Within this study, a commonly applied steel for railway axles, namely EA4T, is investigated as base material. In [33] the crack growth behavior for this steel is presented. The nominal chemical composition and mechanical properties are provided in Tables 1 and 2.

Table 1. Nominal chemical composition of investigated steel material in weight per cent [33].

Steel	C	Si	Mn	Cr	Mo	P	S	Fe
EA4T	0.26	0.29	0.70	1.00	0.20	0.0200	0.007	Balance

Table 2. Nominal mechanical properties of investigated steel material [33].

Steel	f_y (MPa)	f_u (MPa)	A (%)
EA4T	631	789	18.5

As introduced, 1:3 scaled round specimens, which are manufactured from real railway axle blanks are investigated. The testing diameter of the 1:3 scaled specimens measures 55 mm including one semi-elliptical crack, with an initial minimum surface length of about $2s = 4$ mm. The initial crack depth, a, and surface crack length, s, equals a ratio of $a/s = 0.8$. Prior to this initial crack, a semi-elliptical notch with a notch depth of $a = 1$ mm and a surface length of $2s = 2.5$ mm is manufactured by spark eroding. Based on a cyclic load crack initiation procedure, using similar testing conditions as for the subsequent crack growth tests, the described initial crack dimensions are realized. The size of every initial crack is measured at the fracture surface after each crack growth experiment showing a sound reproducibility of the initial crack characteristics.

The crack growth experiments are conducted under rotating bending loading. Details of the testing procedure and the optical surface crack length measurement are provided in [32]. Firstly, constant amplitude load (CAL) tests are performed in order to validate the applicability of the crack growth model for constant load conditions. Secondly, overload tests that consider the varying overload ratios are executed in order to research retardation effects.

Finally, an analytical fatigue crack growth assessment, using the software, Integrity Assessment for Railway Axles (INARA) (Version 19-3-2018_13-47, Materials Center Leoben Forschungs GmbH, Leoben, Austria), which uses a modified NASGRO approach [34] as well as a yield zone concept [35] to cover retardation effects, is performed. The model parameters are based on small-scale SEB tests; hence, a comparison of the model results and the 1:3 experiments demonstrate the applicability of the assessment procedure and the transferability of fracture mechanical parameters. The utilized crack growth model is described in detail in the following.

According to Forman and Mettu [36], the crack growth rate da/dN can be described, based on Equation (1), which is known as the NASGRO equation.

$$\frac{da}{dN} = C \cdot F \cdot \Delta K^m \cdot \frac{\left(1 - \frac{\Delta K_{th}}{\Delta K}\right)^p}{\left(1 - \frac{K_{max}}{K_C}\right)^q} \quad \text{with} \quad K_{max} = \frac{\Delta K}{1 - R} \tag{1}$$

Herein, a is the crack depth, N is the number of load-cycles, F is the crack velocity factor, R is the load stress intensity factor ratio, ΔK is the stress intensity factor range, ΔK_{th} is the threshold stress intensity range, K_C is the fracture toughness, and C, m, p, and q are material constants. The factor F is calculated based on Equation (2) considering the crack opening function f, which is defined as the ratio of the crack opening and maximum value of the stress intensity factor, see [37].

$$F = \left(\frac{1-f}{1-R}\right)^m \quad \text{and} \quad f = \frac{K_{op}}{K_{max}} \tag{2}$$

As shown in Equation (1), the threshold stress intensity range ΔK_{th} acts as one input material parameter for the crack growth assessment. Therefore, not only the threshold of the long crack $\Delta K_{th,lc}$, but also the effective value $\Delta K_{th,eff}$ for physically short cracks need to be considered. The transition from $\Delta K_{th,eff}$ to $\Delta K_{th,lc}$ by a certain value of crack extension Δa is based on crack closure effects [38] and denoted as the crack growth resistance curve for the threshold of the stress intensity range, usually abbreviated as R-curve. Details regarding the determination and limitations are provided in [39]. In [34], the R-curve is defined by Equation (3), where the parameters l_i act as fictitious length scales for the build-up of the different crack closure effects.

$$\Delta K_{th} = \Delta K_{th,eff} + \left(\Delta K_{th,lc} - \Delta K_{th,eff}\right) \cdot \left[1 - \sum_{i=1}^{n} v_i \exp\left(-\frac{\Delta a}{l_i}\right)\right] \quad \text{with} \quad \sum_{i=1}^{n} v_i = 1 \tag{3}$$

Furthermore, an empirical approach for the crack velocity factor F is developed in [34], which additionally considers the R-curve within the evaluation of the factor F, see Equation (4).

$$F = 1 - (1 - F_{lc}) \cdot \left[1 - \exp\left(-\frac{\Delta a}{l_F}\right)\right] \quad \text{with} \quad F_{lc} = \left(\frac{1-f}{1-R}\right)^m \tag{4}$$

Hence, the modified NASGRO equation using Equation (1) considers the R-curve based on Equation (3), and the crack velocity factor F, as given in Equation (4), is applied to assess the crack growth behavior under constant amplitude loading. To additionally cover retardation effects due to overloads under variable amplitude load scenarios, a yield zone concept based on the model by Willenborg [13], Gallagher, and Hughes [16] is presented in [35]. Due to its engineering-feasible applicability, this yield zone concept is used and slightly modified improve the practicability for varying railway axle steel materials, see [35]. Herein, the residual stress intensity factor due to overloads $K_{res,OL}$ can be determined by Equation (5).

$$K_{res,OL} = K_{OL} \cdot K_{max,OL} \cdot \left(1 - \frac{\Delta a}{z_{OL}}\right)^{\gamma_{OL}} - K_{max} \qquad (5)$$

Herein, C_{OL} is a dimensionless constant for the plasticity-induced residual stress intensity factor, $K_{max,OL}$ is the maximum stress intensity factor during the overload, Δa is the crack extension, z_{OL} is the size of the overload influenced zone, γ_{OL} is a material-dependent exponent, and K_{max} is the maximum stress intensity factor of the basic load, see Equation (1).

Originally, the parameter z_{OL} is calculated by $\alpha(K_{max,OL}/\sigma_y)^2$ with α depending on plane stress or strain condition, see [13,16]. In [35], the value z_{OL} is evaluated based on Equation (6), where the parameters L_{OL} and p_{OL} are determined by statistical regression based on experiments with SEB specimens and $\Delta K_{th,0}$ equals the long crack threshold value at $R = 0$.

$$z_{OL} = L_{OL} \cdot (K_{max,OL} - \Delta K_{th,0})^{p_{OL}} \qquad (6)$$

Finally, the crack growth retardation effect, due to overloads is accounted in Equation (1) by considering an effective stress intensity factor ratio R_{eff}, which is calculated by Equation (7). One can see that $K_{res,OL}$ influences R_{eff} and therefore affects the crack growth rate da/dN.

$$R_{eff} = \frac{K_{min} + K_{res,OL}}{K_{max} + K_{res,OL}} \qquad (7)$$

As shown in [35], the parameters of the used retardation model are evaluated based on small-scale SEB tests, with the same base material. In the course of the SEB investigations, it is found that the retardation effect may be over-estimated by the model, leading to significantly low crack growth rates or even no crack propagation at all. This may result in a non-conservative consideration of the retardation effect, which should be avoided in practical application. Hence, an additional factor, denoted as retardation factor RF, is incorporated, which limits the crack growth rate da/dN to a lower boundary value. The factor is defined as the ratio of the reduced crack growth rate after the overload to the crack growth rate, at the base load before the overload. Based on the SEB tests, a value of $RF = 0.10$ is suggested, which equals a maximum decrease of da/dN by the retardation effect down to 10% of the crack growth rate before the overload. The transferability of this RF-value from SEB to round 1:3 scaled railway axle specimens should be validated within this study.

Besides the overload effect, the software package INARA additionally can also take oxide-induced crack closure effects into account. A recently published study [40] presents results of SEB tests, which concludes that the long crack threshold is very sensitive to the influence of oxide debris effects. However, as the focus of the research within this paper is laid on the influence of overloads, due to plasticity-induced retardation effects, the oxide induced crack closure is not considered within this work. Parameters of the applied modified NASGRO model for the investigated steel EA4T are provided in Table 3.

Table 3. Parameters of applied modified NASGRO model for steel EA4T.

C (mm/(MPa$\sqrt{m}$))	m (-)	p (-)	v_1 (-)	v_2 (-)	l_1 (mm)	l_2 (mm)	l_F (mm)	$\Delta K_{th,eff}$ (MPa$\sqrt{m}$)	$\Delta K_{th,0}$ (MPa$\sqrt{m}$)
1.92×10^{-8}	2.64	0.32	0.43	0.57	0.41e-3	1.75	0.01	2.00	7.35

Regarding the retardation model, values of $C_{OL} = 1.0$, $\gamma_{OL} = 0.37$, $p_{OL} = 2.72$, and $L_{OL} = 7.62 \times 10^{-4}$ mm are used, for details see [35]. In accordance with a preceding study [41] that focuses on the constant crack growth behavior of another commonly used steel material for railway axles, namely EA1N, the stress intensity factor for the semi-elliptical crack in round bars is analytically calculated according to [42], and furthermore summarized in [43]. As aforementioned, the surface crack length is optically measured during the experiments [32]; hence, in the following, all test results, as well as crack propagation parameters are related to the crack extension at the surface. Due to the cut-out of the specimens from real railway axle blanks, minor residual stresses up to 20 MPa are still measured, see [41]. As it is highlighted in [41] that these comparably minor residual stresses significantly affect the crack growth characteristics, the accordant residual stress values are considered within this study to properly assess the crack propagation. Further details are given in [41].

3. Results

3.1. Constant Amplitude Tests

At first, constant amplitude load tests (CAL) are performed to validate the applicability of the utilized crack propagation model under constant loads. Figure 1 shows the results of CAL test #1, which is tested at a nominal bending stress amplitude of σ_a = 100 MPa. During the experiment, the surface crack length, $2s$, is optically measured and the crack propagation test is stopped at a final crack length of $2s$~18 mm. Utilizing the crack length $2s$ versus the accordant number of load-cycles, N, in Figure 1a, the crack propagation rate $d(2s)/dN$ is computed. The corresponding surface stress intensity factor range ΔK_S is calculated based on the procedure in [42], thereby enabling the representation of the $d(2s)/dN$ vs. ΔK_S diagram, as depicted in Figure 1b, for further comparison with the crack propagation model.

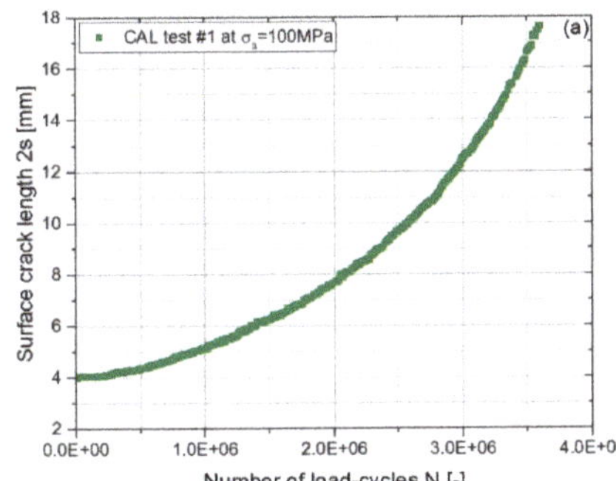

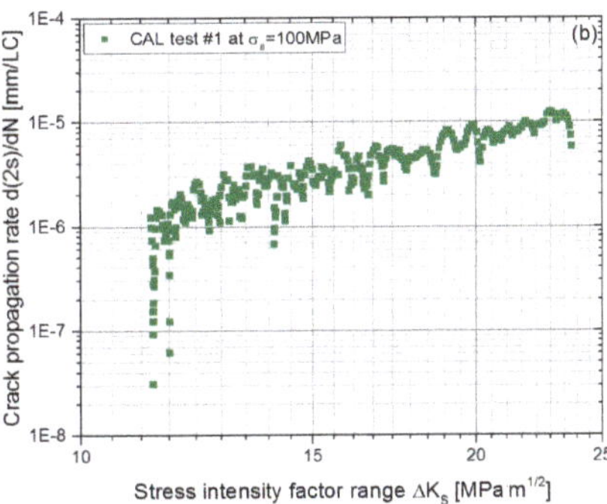

Figure 1. Results of amplitude load tests (CAL) test #1 at load of σ_a = 100 MPa (**a**) surface crack length vs. load-cycles; (**b**) crack propagation rate versus stress intensity factor.

At the beginning of the experiment, decreased crack growth rates down to a value of about 1e-7 mm per load-cycle (LC) are observable, which occur due to the short crack effect [34,39], as well as the prior crack initiation procedure. However, after this initiation and short crack phase, the crack constantly grows leading to a $d(2s)/dN$ vs. ΔK_S curve as presented. The fracture surface of CAL test #1 is illustrated in Figure 2. One can clearly see the initial starting notch at the top center of the picture as well as the further crack propagation area. At the final surface crack length of $2s$~18 mm, the crack depth exhibits a~8 mm leading to a final ratio of a/s~0.9.

In Figure 3, the results of CAL test #2, which is tested at an increased nominal bending stress amplitude of σ_a = 150 MPa, is demonstrated. Starting from the same crack length of $2s$ = 4 mm as for CAL test #1, the total lifetime until the final crack length of $2s$~18 mm is only about 1×10^6 load-cycles, due to the increased bending load, see Figure 3a. Again, the crack propagation rate $d(2s)/dN$ versus the surface stress intensity factor range ΔK_S is evaluated, shown in Figure 3b.

Based on the described modified NASGRO equation, the constant crack propagation model is applied and the results are compared to the results of CAL test #1 and #2. Cyclic crack resistance (R-) curves as well as the constant long crack growth behavior for the investigated EA4T steel are presented in [27], which act as basis for the crack growth assessment in this work. A comparison of the crack propagation model with the CAL tests is demonstrated in Figure 4. It is shown that the model fits well to both CAL crack propagation tests with a somewhat conservative assessment for lower stress intensity factor ranges. However, these results prove the transferability of the model parameters evaluated by small-scale SEB specimens to semi-elliptically cracked round bars. Further details of the transferability and used parameters are provided in [41].

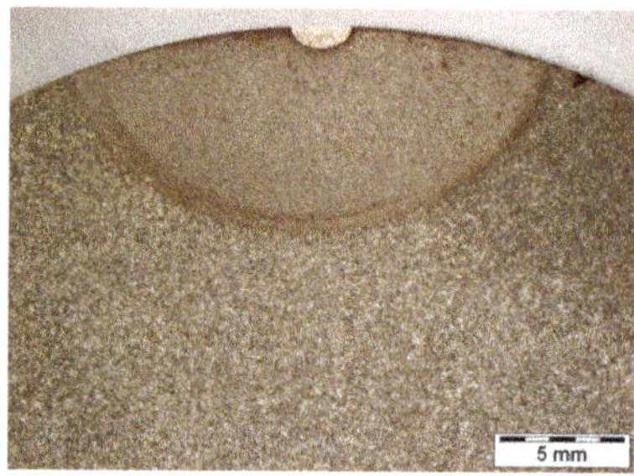

Figure 2. Representation of fracture surface for specimen utilized within CAL test #1.

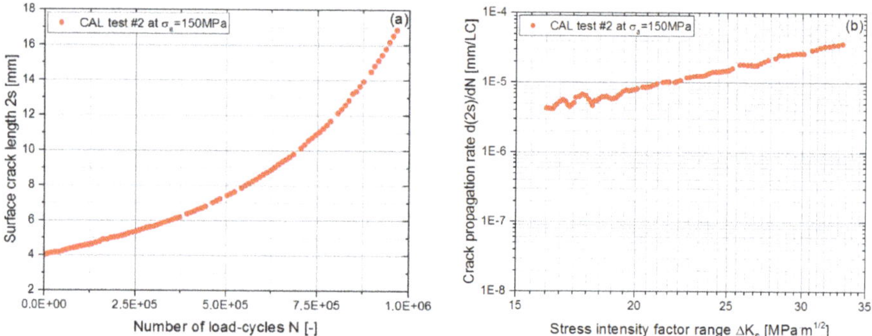

Figure 3. Results of CAL test #2 at load of σ_a = 150 MPa (**a**) surface crack length vs. load-cycles; (**b**) crack propagation rate vs. stress intensity factor.

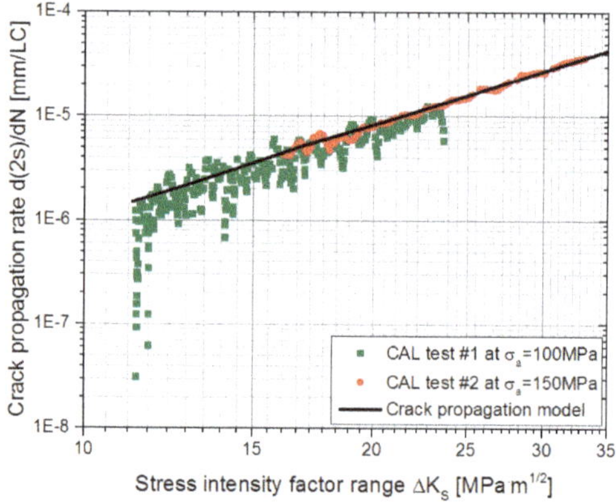

Figure 4. Comparison of crack propagation model with results of CAL test #1 and #2.

3.2. Overload Tests

3.2.1. Overload Ratio R_{OL} = 2.0

Second, crack propagation tests, including overloads are performed in order to validate the applicability of the model to cover retardation effects. In Figure 5, the results of Overload test #1 applying ten overloads under the same load stress ratio of $R = -1$ with an overload ratio of $R_{OL} = 2.0$ is illustrated. The surface crack length $2s$ over load-cycles N in Figure 5a shows that the overload is applied at $2s{\sim}9.4$ mm, which equals a surface stress intensity factor range of $\Delta K_S{\sim}17$ MPa·m$^{1/2}$ under the base load bending stress of $\sigma_a = 100$ MPa.

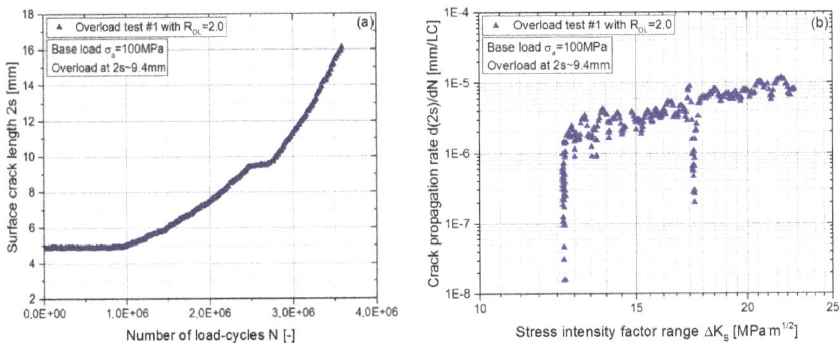

Figure 5. Results of Overload test #1 with $R_{OL} = 2.0$ (**a**) surface crack length versus load-cycles; (**b**) crack propagation rate vs. stress intensity factor.

The retardation effect is clearly observable within the $d(2s)/dN$ versus ΔK_S diagram in Figure 5b, whereas the crack growth rate after the overload at $\Delta K_S{\sim}17$ MPa·m$^{1/2}$ is significantly decreased. However, after a certain number of further load-cycles at the base load, denoted as delay cycles N_d [44], the effect of the overload is passed and the crack propagation rate proceeds in accordance with the constant load tests. For Overload test #1, the delay cycle number is evaluated to $N_d{\sim}2.6 \times 10^5$, which proves the beneficial retardation effect.

The specimen's fracture surface of Overload test #1 is depicted in Figure 6. In accordance with the CAL tests, the initial starting notch is again observable at the top middle. Again, the crack initiation is visible due to minor crack propagation rates at the beginning of the experiment, compare to Figure 5b, which merges into the crack growth regime. Reaching a surface crack length of $2s{\sim}9.4$ mm, the applied overload is clearly detectable in the fracture surface. At the end of the experiment, the final a/s-ratio equals about a value of 0.9, which is in accordance to the CAL tests.

Figure 6. Representation of fracture surface for specimen utilized within Overload test #1.

Similar to the first overload test, Figure 7 shows the results of the Overload test #2, applying the identical overload ratio of $R_{OL} = 2.0$ under the same testing conditions. Again, the retardation effect is pronounced, leading to a delay cycle number of $N_d \sim 1.7 \times 10^5$. Compared with the Overload test #1 this value is reduced; however, it is still beneficial influence as the overload is recognizable.

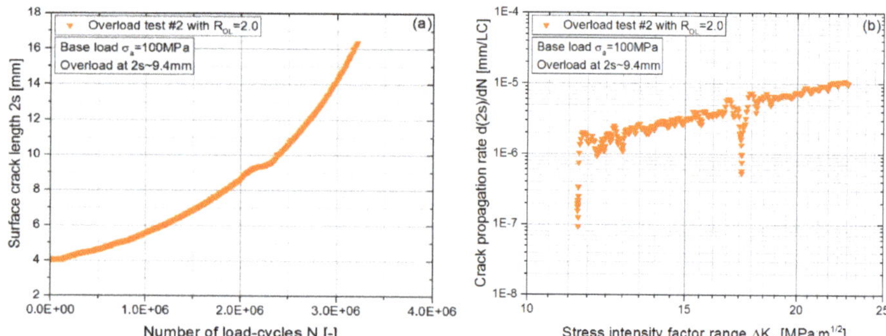

Figure 7. Results of Overload test #2 with $R_{OL} = 2.0$ (**a**) surface crack length versus load-cycles; (**b**) crack propagation rate vs. stress intensity factor.

To assess the overload-induced retardation effect, the previously described procedure, using a modified method based on the approach by Willenborg et al. is applied. Values for the investigated steel EA4T are evaluated on the basis of SEB tests and are provided in Table 3 on the basis of [35]. These parameters are used to estimate the overload effect for the semi-elliptically cracked round bars within this study; therefore, validating the transferability of the model parameters from small-scale SEB to 1:3 scaled railway axle specimen tests, incorporating varying global specimen geometry, as well as different shapes of the crack front.

As previously described, within the crack growth assessment the additional retardation factor *RF* is included, which defines the maximum decrease of the crack propagation rate, due to the overload compared to the prior base load. A factor of *RF* = 0.10 means that subsequent to the overload, the retardation effect can decrease the crack propagation rate to a minimum of 10% of the value at the base load, before the overload. In this work, two different *RF*-values in particular, *RF* = 0.10 and *RF* = 0.05, are analyzed to highlight the impact of the retardation factor on the overload effect. As stated, a value of *RF* = 0.10 is suggested, based on preliminary performed SEB overload tests in [35]. The transferability of this value *RF* = 0.10 and additionally the effect of using *RF* = 0.05, which enables a more pronounced retardation effect within the model, is studied.

A comparison of the crack propagation model with the results of the Overload test #1 and #2, both with $R_{OL} = 2.0$ is shown in Figure 8. In general, the results reveal a sound agreement between the model and the experiments. The parameter set, considering *RF* = 0.05 exhibits a greater decrease of the retardation-affected crack growth rate, leading to a delay cycle number of $N_d \sim 3.78 \times 10^5$ compared to *RF* = 0.10 with $N_d \sim 2.1 \times 10^5$ as highlighted in the preceding paragraph. However, the applied model seems to cover both the constant amplitude as well as the overload-affected region well. A further discussion comparing the delay cycle number N_d of the model to the experiments is given in Section 4.

Besides the retardation effect, a different crack growth behavior, at the beginning of both tests, can be observed. As aforementioned within the CAL tests, the decreased crack propagation rates at the beginning are a result of the short crack effect as well as the prior crack initiation procedure. The deviation in this case can be primarily explained by a varying initial surface crack length *2s* between both tests. However, the same load stress amplitude is used and the overload is applied at the same surface crack length, thereby ensuring a sound comparison of both overload test results.

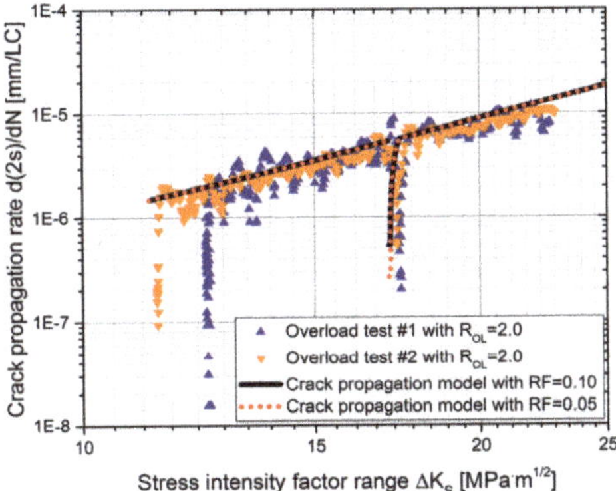

Figure 8. Comparison of crack propagation model with results of Overload test #1 and #2.

3.2.2. Overload Ratio R_{OL} = 2.5

Similar to the Overload tests #1 and #2 with R_{OL} = 2.0, two further experiments with R_{OL} = 2.5, denoted as Overload tests #3 and #4, are conducted. The test results are depicted in Figures 9 and 10 respectively, which highlight a pronounced retardation effect in both cases. The delay cycle number of Overload test #3 is evaluated to $N_d = 4.7 \times 10^5$, and for Overload test #4 to $N_d = 8.5 \times 10^5$. On average, this equals an increase in N_d from R_{OL} = 2.0 to R_{OL} = 2.5 by about a factor of three, proving the significant impact of the overload ratio R_{OL} on the retardation effect.

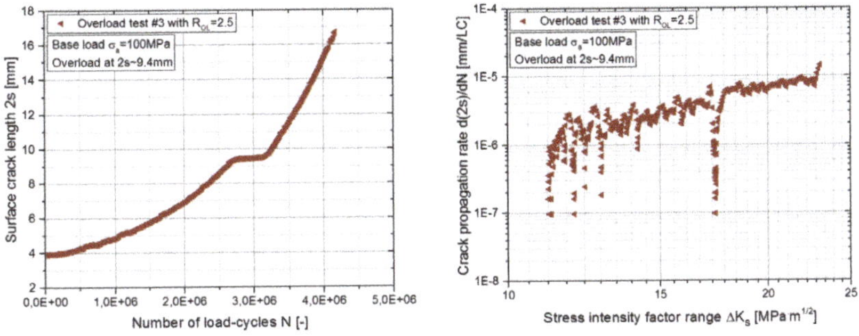

Figure 9. Results of Overload test #3 with R_{OL} = 2.5 (**a**) surface crack length versus load-cycles; (**b**) crack propagation rate vs. stress intensity factor.

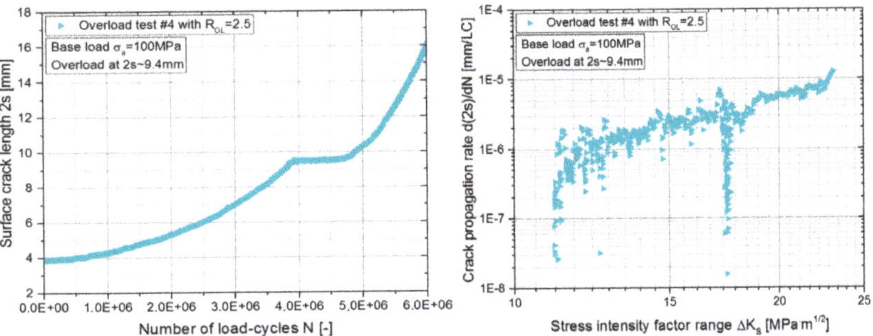

Figure 10. Results of Overload test #4 with R_{OL} = 2.5 (**a**) surface crack length versus load-cycles; (**b**) crack propagation rate vs. stress intensity factor.

As presented for $R_{OL} = 2.0$ in Figure 8 before, the Overload tests #3 and #4 with $R_{OL} = 2.5$ are again compared to the crack propagation model considering $RF = 0.10$ and $RF = 0.05$, see Figure 11. Here, the model again fits well to both experiments, whereas the retardation factor of $RF = 0.10$ leads to a reduced pronounced overload effect, with a final delay cycle number of $N_d \sim 6.7 \times 10^5$ compared with the value of $RF = 0.05$ leading to $N_d \sim 1.6 \times 10^6$. A comparison of the delay cycles N_d of the model and the experiments is shown in the next section.

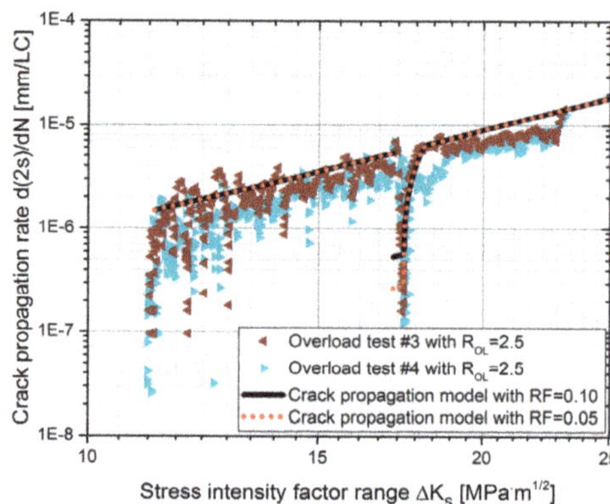

Figure 11. Comparison of crack propagation model with results of Overload test #3 and #4.

4. Discussion

As the focus of this paper is laid on the applicability of the presented crack growth assessment for overload-induced retardation effects, this section compares the overload tests with the crack propagation model in terms of the evaluated delay cycle numbers N_d. Table 4 summarizes the values of N_d for the Overload tests #1 and #2 with $R_{OL} = 2.0$ with the results of the model, by considering a retardation factor $RF = 0.10$ and $RF = 0.05$. On average, a delay cycle number of $N_d \sim 2.2 \times 10^5$ is evaluated for the experiments, which equals well the value of 2.1×10^5 of the model using $RF = 0.10$. Applying a factor of $RF = 0.05$, a significantly increased pronounced overload effect occurs, which leads to a non-conservative delay cycle number of 3.8×10^5.

Table 4. Delay cycles N_d by experiment and model for overload tests with $R_{OL} = 2.0$.

Experiment	N_d by Experiment (-)	N_d by Crack Propagation Model (-)
Overload test #1	2.6×10^5	
Overload test #2	1.7×10^5	2.1×10^5 ($RF = 0.10$) and 3.8×10^5 ($RF = 0.05$)

A similar analysis for the Overload tests #3 and #4 with $R_{OL} = 2.5$ is provided in Table 5. There, a mean value of of $N_d \sim 6.6 \times 10^5$ is evaluated for the experiments, which again matches well to the value of 6.7×10^5 of the model using $RF = 0.10$. Considering $RF = 0.05$ leads to a non-conservative assessment with a delay cycle number of 1.6×10^6 as shown for $R_{OL} = 2.0$.

Table 5. Delay cycles N_d by experiment and model for overload tests with $R_{OL} = 2.5$.

Experiment	N_d by Experiment (-)	N_d by Crack Propagation Model (-)
Overload test #3	4.7×10^5	
Overload test #4	8.5×10^5	6.7×10^5 ($RF = 0.10$) and 1.6×10^6 ($RF = 0.05$)

As shown in Table 5, a comparably increased deviation of N_d between both experiments can be observed. Thereby, the N_d-values are by trend in line with the crack growth behavior at the constant base load, whereby test #3, which exhibits a minor value of N_d due to the overload, reveals a comparably increased crack propagation rate compared to test #4.

5. Conclusions

Based on the investigations in this work, the following scientific conclusions can be drawn:

- Retardation effects, due to the overloads, significantly affect the crack growth rate leading to an enhancement of the lifetime. Considering the presented test results at overload ratios of $R_{OL} = 2.0$ and $R_{OL} = 2.5$, the influence is more pronounced at higher R_{OL}-values.
- The presented crack propagation model based on a modified NASGRO equation and considering the approach by Willenborg, Gallagher, and Hughes to cover retardation effects fits well with the conducted 1:3 round specimen overload tests. The additionally introduced retardation factor RF, which defines the maximum decrease of the crack propagation rate due to overloads, seems to exhibit a remarkable influence on the delay cycle number N_d. In this study, the suggested value of $RF = 0.10$, which is evaluated based on preceding SEB tests, maintains a sound applicability.
- As all model parameters are evaluated on the basis of small-scale SEB tests, the transferability of these values, by considering the effect of specimen size, geometry, as well as shape of the crack front, is validated based on the results in this study.

Further work will focus on the interaction of the presented overload with oxide-induced retardation effects [40], as well as the influence of variable amplitude [45], and multiaxial [46] loads.

Author Contributions: Conceptualization, M.L. and D.S.; methodology, M.L.; software, M.L., D.S., J.M., and H.-P.G.; validation, M.L. and D.S.; data curation, D.S. and J.M.; writing—original draft preparation, M.L.; writing—review and editing, D.S., J.M., H.-P.G., and R.P.

Funding: Scientific support was given within the framework of the COMET K2-Programme, whereby the Austrian Federal Government represented by Österreichische Forschungsförderungsgesellschaft mbH and the Styrian and the Tyrolean Provincial Government, represented by Steirische Wirtschaftsförderungsgesellschaft mbH and Standortagentur Tirol, is gratefully acknowledged.

Conflicts of Interest: The authors declare no conflict of interest.

References

1. Palmgren, A. Die Lebensdauer von Kugellagern. *Zeitschrift des Vereins Deutscher Ingenieure* **1924**, *68*, 339–341.
2. Miner, M.A. Cumulative Damage in Fatigue. *J. Appl. Mech. Trans. ASME* **1945**, *12*, A159–A164.
3. Ciavarella, M.; D'antuono, P.; Papangelo, A. On the connection between Palmgren-Miner rule and crack propagation laws. *Fatigue Fract. Eng. Mater. Struct.* **2018**, *41*, 1469–1475. [CrossRef]
4. Wei, R.P.; Shih, T.T. Delay in fatigue crack growth. *Int. J. Fract.* **1974**, *10*, 77–85. [CrossRef]
5. Carlson, R.L.; Kardomateas, G.A.; Bates, P.R. The effects of overloads in fatigue crack growth. *Int. J. Fatigue* **1991**, *13*, 453–460. [CrossRef]
6. Schijve, J. *Fatigue Crack Propagation in Light Alloy Sheet Materials and Structures*; NRL Report MP 195; Elsevier: Amsterdam, The Netherlands, 1960.
7. Lankford, J.; Davidson, D.L. The effect of overloads upon fatigue crack tip opening displacement and crack tip opening/closing loads in aluminium alloys. In *Advances in Fracture Research, (Fracture 81), Proceedings of the 5th International Conference on Fracture (ICF5), Cannes, France, 29 March–3 April 1981*; François, D., Ed.; Pergamon: Oxford, NY, USA, 1982; pp. 899–906.
8. Elber, W. The Significance of Fatigue Crack Closure. In *Damage Tolerance in Aircraft Structures: A Symposium Presented at the Seventy-Third Annual Meeting American Society for Testing and Materials, Toronto, ON, Canada, 21–26 June 1970*; Rosenfeld, M.S., Ed.; American Society for Testing and Materials: Philadelphia, PA, USA, 1971.
9. Jones, R.E. Fatigue crack growth retardation after single cycle peak overload Ti-6Al-4V titanium alloy. *Eng. Fract. Mech.* **1973**, *5*, 585–604. [CrossRef]
10. Christensen, R.H. *Metal Fatigue*; McGraw-Hill: New York, NY, USA, 1959.

11. Ricardo, L.C.H.; Miranda, C.A.J. Crack simulation models in variable amplitude loading—A review. *Frattura ed Integrità Strutturale* **2016**, *10*, 456–471. [CrossRef]
12. Wheeler, O.E. Spectrum loading and crack growth. *J. Basic Eng. Trans. ASME* **1972**, *94*, 181–186. [CrossRef]
13. Willenborg, J.; Engle, R.M.; Wood, H.A.; Air Force Flight Dynamics Lab Wright-Patterson AFB OH. *A Crack Growth Retardation Model Using an Effective Stress Concept*; Defense Technical Information Center: Fort Belvoir, VA, USA, 1971.
14. Porter, T.R. Method of analysis and prediction of variable amplitude fatigue crack growth. *Eng. Fract. Mech.* **1972**, *4*, 717–736. [CrossRef]
15. Gray, T.D.; Gallagher, J.P. Predicting Fatigue Crack Retardation Following a Single Overload Using a Modified Wheeler Model. In *Mechanics of Crack Growth*; Rice, J.R., Ed.; American Society for Testing & Materials: West Conshohocken, PA, USA, 1976.
16. Gallagher, J.P.; Hughes, T.F. Influence of Yield Strength on Overload Affected Fatigue Crack Growth Behavior in 4340 Steel. 1974. Available online: http://www.dtic.mil/dtic/tr/fulltext/u2/787655.pdf (accessed on 29 January 2019).
17. Johnson, W.S. Multi-Parameter Yield Zone Model for Predicting Spectrum Crack Growth. In *Methods and Models for Predicting Fatigue Crack Growth under Random Loading*; Chang, J.B., Hudson, C.M., Eds.; American Society for Testing and Materials: Philadelphia, PA, USA, 1981.
18. Chang, J.B.; Hiyama, R.M.; Szamossi, M. *Improved Methods for Predicting Spectrum Loadings Effects*; AFFDL-TR-79-3036; AFFDL: Air Force, OH, USA, 1984.
19. Bell, P.D.; Creager, M. *Crack Growth Analyses for Arbitrary Spectrum Loading*; AFFDL-TR-74-129; AFFDL: Air Force, OH, USA, 1974.
20. Newman, J.C. A Finite-Element Analysis of Fatigue Crack Closure. In *Mechanics of Crack Growth*; Rice, J.R., Ed.; American Society for Testing & Materials: West Conshohocken, PA, USA, 1976.
21. Dill, H.D.; Saff, C.R. Spectrum Crack Growth Prediction Method Based on Crack Surface Displacement and Contact Analyses. In *Fatigue Crack Growth under Spectrum Loads*; Wei, R.P., Ed.; American Society for Testing & Materials: West Conshohocken, PA, USA, 1976.
22. Kanninen, M.F.; Atkinson, C.; Feddersen, C.E. *A Fatigue Crack Growth Analysis Method Based on a Single Representation of Crack Tip Plasticity*; ASTM Special Technical Publication; ASTM: West Conshohocken, PA, USA, 1977.
23. Budiansky, B.; Hutchinson, J.W. Analysis of Closure in Fatigue Crack Growth. *J. Appl. Mech.* **1978**, *45*, 267–276. [CrossRef]
24. De Koning, A.U. A Simple Crack Closure Model for Prediction of Fatigue Crack Growth Rates Under Variable-Amplitude Loading. In *Fracture Mechanics: Proceedings of the Thirteenth National Symposium on Fracture Mechanics*; Roberts, R., Ed.; American Society for Testing and Materials: Philadelphia, PA, USA, 1981.
25. Pereira, M.V.S.; Darwish, F.A.I.; Camarão, A.F.; Motta, S.H. On the prediction of fatigue crack retardation using Wheeler and Willenborg models. *Mater. Res.* **2007**, *10*, 101–107. [CrossRef]
26. Schijve, J. *Fatigue Crack Growth Predictions for Variable-Amplitude and Spectrum Loading*; LR-526; Delft University of Technology: Delft, The Netherlands, 1987.
27. Gänser, H.-P.; Maierhofer, J.; Tichy, R.; Zivkovic, I.; Pippan, R.; Luke, M.; Varfolomeev, I. Damage tolerance of railway axles—The issue of transferability revisited. *Int. J. Fatigue* **2016**, *86*, 52–57. [CrossRef]
28. Traupe, M.; Jenne, S.; Lütkepohl, K.; Varfolomeev, I. Experimental validation of inspection intervals for railway axles accompanying the engineering process. *Int. J. Fatigue* **2016**, *86*, 44–51. [CrossRef]
29. Zerbst, U.; Schödel, M.; Beier, H.T. Parameters affecting the damage tolerance behaviour of railway axles. *Eng. Fract. Mech.* **2011**, *78*, 793–809. [CrossRef]
30. Zerbst, U.; Beretta, S.; Köhler, G.; Lawton, A.; Vormwald, M.; Beier, H.T.; Klinger, C.; Černý, I.; Rudlin, J.; Heckel, T.; et al. Safe life and damage tolerance aspects of railway axles—A review. *Eng. Fract. Mech.* **2013**, *98*, 214–271. [CrossRef]
31. Pokorný, P.; Hutař, P.; Náhlík, L. Residual fatigue lifetime estimation of railway axles for various loading spectra. *Theor. Appl. Fract. Mech.* **2016**, *82*, 25–32. [CrossRef]
32. Simunek, D.; Leitner, M.; Maierhofer, J.; Gänser, H.-P. Crack growth under constant amplitude loading and overload effects in 1:3 scale specimens. *Procedia Struct. Integr.* **2017**, *4*, 27–34. [CrossRef]
33. Linhart, V.; Černý, I. An effect of strength of railway axle steels on fatigue resistance under press fit. *Eng. Fract. Mech.* **2011**, *78*, 731–741. [CrossRef]

34. Maierhofer, J.; Pippan, R.; Gänser, H.-P. Modified NASGRO equation for physically short cracks. *Int. J. Fatigue* **2014**, *59*, 200–207. [CrossRef]

35. Maierhofer, J.; Gänser, H.-P.; Pippan, R. Crack closure and retardation effects—Experiments and modelling. *Procedia Struct. Integr.* **2017**, *4*, 19–26. [CrossRef]

36. Forman, R.; Mettu, S.R. *Behavior of Surface and Corner Cracks Subjected to Tensile and Bending Loads in Ti-6A1-4V Alloy*; NASA Technical Memorandum 102165: Houston, TX, USA, 1990.

37. Newman, J.C. A crack opening stress equation for fatigue crack growth. *Int. J. Fract.* **1984**, *24*, R131–R135. [CrossRef]

38. Pippan, R.; Hohenwarter, A. Fatigue crack closure: A review of the physical phenomena. *Fatigue Fract. Eng. Mater. Struct.* **2017**, *40*, 471–495. [CrossRef] [PubMed]

39. Maierhofer, J.; Kolitsch, S.; Pippan, R.; Gänser, H.-P.; Madia, M.; Zerbst, U. The cyclic R-curve—Determination, problems, limitations and application. *Eng. Fract. Mech.* **2018**, *198*, 45–64. [CrossRef]

40. Maierhofer, J.; Simunek, D.; Gänser, H.-P.; Pippan, R. Oxide induced crack closure in the near threshold regime: The effect of oxide debris release. *Int. J. Fatigue* **2018**, *117*, 21–26. [CrossRef]

41. Simunek, D.; Leitner, M.; Maierhofer, J.; Gänser, H.-P. Analytical and Numerical Crack Growth Analysis of 1:3 Scaled Railway Axle Specimens. *Metals* **2019**, in review.

42. Varfolomeev, I.; Burdack, M.; Luke, M. *Fracture Mechanics as a Tool for Specifying Inspection Intervals of Railway Axles (Part 2)*; DVM: Dresden, Germany, 2007.

43. Madia, M.; Beretta, S.; Schödel, M.; Zerbst, U.; Luke, M.; Varfolomeev, I. Stress intensity factor solutions for cracks in railway axles. *Eng. Fract. Mech.* **2011**, *78*, 764–792. [CrossRef]

44. Ishihara, S.; McEvily, A.; Goshima, T.; Nishino, S.; Sato, M. The effect of the R value on the number of delay cycles following an overload. *Int. J. Fatigue* **2008**, *30*, 1737–1742. [CrossRef]

45. Simunek, D.; Leitner, M.; Maierhofer, J.; Gänser, H.-P. Fatigue Crack Growth Under Constant and Variable Amplitude Loading at Semi-elliptical and V-notched Steel Specimens. *Procedia Eng.* **2015**, *133*, 348–361. [CrossRef]

46. Leitner, M.; Tuncali, Z.; Steiner, R.; Grün, F. Multiaxial fatigue strength assessment of electroslag remelted 50CrMo4 steel crankshafts. *Int. J. Fatigue* **2017**, *100*, 159–175. [CrossRef]

Article

Fatigue Limit Improvement and Rendering Defects Harmless by Needle Peening for High Tensile Steel Welded Joint

Ryutaro Fueki [1], Koji Takahashi [2,*] and Mitsuru Handa [3]

[1] Graduate School of Engineering, Yokohama National University, 79-5, Tokiwadai, Hodogaya-ku, Yokohama-city, Kanagawa 240-8501, Japan; fueki-ryuutaro-pv@ynu.jp

[2] Faculty of Engineering, Yokohama National University, Yokohama-city, Kanagawa 240-8501, Japan

[3] Technology Development Group, Toyo Seiko Co., Ltd., Yatomi-city, Aichi 490-1412, Japan; m_handa@toyoseiko.co.jp

* Correspondence: takahashi-koji-ph@ynu.ac.jp; Tel.:+81-45-339-4017

Received: 11 January 2019; Accepted: 26 January 2019; Published: 28 January 2019

Abstract: The effects of needle peening (NP) on the bending fatigue limit of a high tensile steel (HTS) HT780 (JIS-SHY685)-welded joint containing an artificial semicircular slit on the weld toe were investigated. Three-point bending fatigue tests were conducted at a stress ratio of $R = 0.05$ for NP-treated welded specimens with and without a slit. The fatigue limits of all specimens increased by 9–200% due to the NP treatment. Furthermore, NP-treated specimens with slit depths of $a = 1.0$ mm exhibited high fatigue limits that were equal to those of NP-treated specimens without a slit. Therefore, a semicircular slit of less than $a = 1.0$ mm could be rendered harmless through NP treatment. This result indicates that the reliability of HTS-welded joints can be significantly improved via NP for surface defects with depths that are less than 1 mm, which are not detected through non-destructive inspection (NDI). Therefore, the problem regarding the reliability of HTS-welded joints that restricts the industrial utilization of HTS can be solved by performing both NDI and NP. The dominant factor that contributed to the improvement of the fatigue limit and increase in the acceptable defect size was the introduction of large and deep compressive residual stress with non-propagating cracks.

Keywords: welded joint; needle peening; compressive residual stress; surface defect; fatigue limit

---

1. Introduction

Recently, the requirements of high tensile steel (HTS) have increased due to the growing demand to construct larger steel structures and reduce weight during steel transportation. However, utilizing welded HTS structures is difficult because the fatigue strength of the welded joints of HTS is at the same level of that of mild steel due to tensile residual stresses and stress concentration.

Recent studies showed that the fatigue strength of welded joints could be improved via weld toe treatments, such as needle peening (NP) [1,2] and high-frequency mechanical impact (HFMI) [3–8]. NP and HFMI introduce beneficial compressive residual stresses in the surface layer around the weld toe by repeatedly striking the surface with a needle pin, as shown in Figure 1. In addition, applying these methods results in work hardening and reduces the stress concentration at the weld toe.

Another problem of HTS is its low defect tolerance as the fatigue strength of HTS significantly decreases even if the defect size is small. Therefore, it is difficult to ensure the reliability of a HTS-welded joint even if non-destructive inspection (NDI) is conducted as small defects below the detection limit of NDI cannot be identified.

Research on the effectiveness of the peening treatment in improving the reliability of metal materials has been conducted based on the idea that compressive residual stress introduced by peening could restrain crack opening and render the crack harmless [9–16]. For example, Takahashi et al. [9] reported that the fatigue limit of a spring steel JIS-SUP9A specimen, which contained crack-like semicircular slits with a depth of 0.2 mm that decreased the fatigue limit by 50%, could be rendered harmless through shot peening. Furthermore, Houjou et al. [13] reported that the acceptable semicircular slit depths that could be rendered harmless after NP were 1.0 and 1.2 mm for stainless steel JIS-SUS316 and low-carbon steel JIS-SM490A welded joints, respectively. These results indicate that surface defects that are smaller than the detection limit of NDI, which was reported to be approximately 1-mm deep using a typical eddy current flaw detection method [17], can be rendered harmless by NP.

If the surface defects below the detection limit of NDI in a HTS-welded joint could be rendered harmless by NP, the reliability of the welded joint could be significantly improved and the industrial use of HTS would be expanded. Therefore, the present study involved clarifying the surface defect size in a HTS-welded joint that can be rendered harmless by NP.

Specifically, fatigue tests were conducted for both NP-treated and untreated HTS-welded specimens that contained crack-like semicircular slits at the weld toe. The fatigue limit of metals changes depending on residual stress, hardness, metal microstructure and stress concentrations. Therefore, the changes of these factors after NP were investigated to better understand the fatigue test results. Furthermore, the presence of non-propagating cracks was investigated on NP-treated welded specimens with a semicircular slit tested under the fatigue limit to ascertain whether an acceptable defect size could be determined through a crack initiation condition from the slit or a threshold condition for the crack propagation.

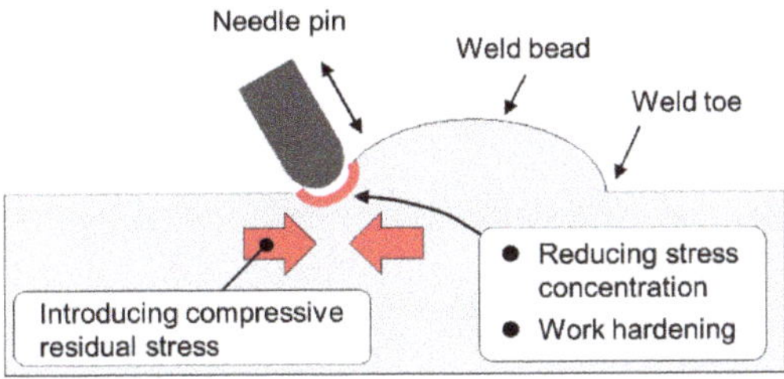

Figure 1. Schematic view of needle peening (NP) at the weld toe. NP introduces beneficial compressive residual stresses in the surface layer around the weld toe by repeatedly striking the surface with a needle pin.

2. Materials and Methods

2.1. Test Material and Specimens

The test material was heat-treated HTS HT780 (JIS-SHY685). Tables 1 and 2 show its chemical composition and mechanical properties, respectively.

The test material was supplied in the form of 7-mm-thick plates, as shown in Figure 2a. Tungsten inert gas welding was performed using a welding rod on the center of the narrow side of the plate. An automatic welding machine (Kobe Steel, Ltd., Kobe, Japan) was used to make the shape and dimensions of the weld bead as constant as possible. Table 3 presents the details of the welding conditions. The welded plate was cut to prepare the test specimens such that the longitudinal direction of the specimen corresponded to the rolling direction of the plate. Thirteen welded specimens were obtained from a welded plate. The shape and dimensions of the test specimen are shown in Figure 2b.

Table 1. Chemical composition of HT780 (mass %).

C	Si	Mn	P	S	Ni	Cr	Mo	Nb	B
0.14	0.35	1.18	0.005	0.001	0.01	0.09	0.12	0.02	0.001

Table 2. Mechanical proparties of HT780.

Yield Stress (MPa)	Ultimate Tensile Stress (MPa)	Vickers Hardness *HV*
822	839	267

Table 3. Welding conditions.

Parameters	Conditions
Number of layers	1
Number of passes	1
Welding position	Flat
Diameter of the welding rod (mm)	Φ 2.4
Current (A)	180
Voltage (V)	9.7
Welding speed (cm/min)	12

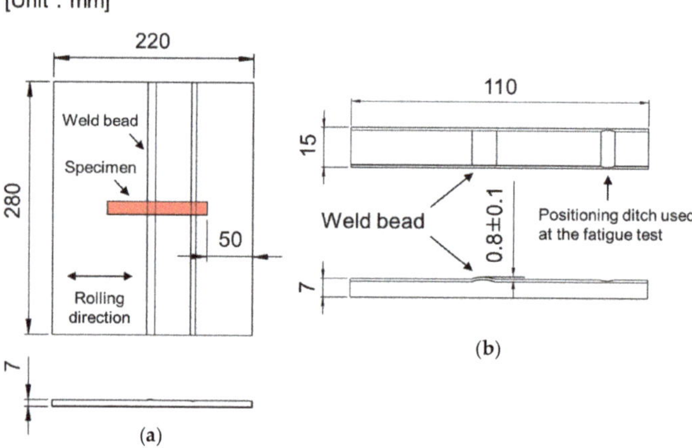

Figure 2. Shape and dimensions of (**a**) the welded plate and (**b**) the test specimen. The test material was supplied in the form of 7-mm-thick plates. The welded plate was cut to prepare the test specimens such that the longitudinal direction of the specimen corresponded to the rolling direction of the plate.

2.2. Needle Peening and the Introduction of the Surface Defect

Figure 3 shows the flow chart of the specimen preparation procedure and the types of specimens used for the fatigue test. Four groups of specimens were prepared, which were namely the welded specimen (W), NP-treated welded specimen (WN), welded specimen with a semicircular slit at the weld toe (WS) and NP-treated welded specimen with a semicircular slit at the weld toe (WNS).

NP was performed on both sides of the weld toe of welded specimens using a portable pneumatic needle-peening (PPP) unit. The PPP system manufactured by Toyo Seiko Co., Ltd., Yatomi, Japan, is a lightweight and hand-portable piece of equipment, which simplifies the application of this technique to various welded structures [2]. The peening conditions are shown in Table 4.

A semicircular slit similar to a surface crack was introduced after NP treatment on WNS to avoid the deformation of the slit by NP. Electronic discharge machining (Sodick Co.,Ltd., Yokohama, Japan) was used to introduce the semicircular slit. Figure 4 shows the slit introduction position, slit shape and dimensions. The slit was introduced within 0.1 mm of the weld toe of WS (Figure 4a). The machining position of the slit of WNS was at the bottom of the dent created by NP, as shown in Figure 4b. The dimensions of the slits were divided into two groups. As shown in Figure 4c, the diameter ($2a$) to depth (a) ratios in mm were 2.0/1.0 and 3.0/1.5. The slit size was determined based on the results of previous studies [13,15], which found the acceptable defect size after NP. The aperture width of the slit was approximately 0.1 mm. The depth of the slit is appended to the abbreviation for WS and WNS. For example, WNS1.0 denotes the NP-treated welded specimen with a 1.0-mm-deep semicircular slit at the weld toe.

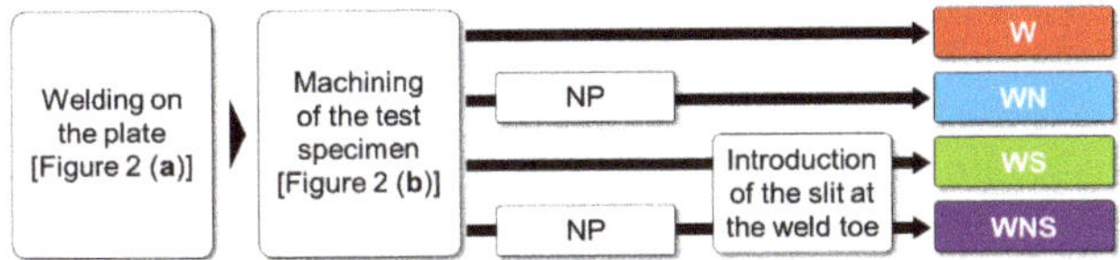

Figure 3. Flow chart of specimen preparation and types of specimens. Four groups of specimens were prepared, which were namely the welded specimen (W), NP-treated welded specimen (WN), welded specimen with a semicircular slit at the weld toe (WS) and NP-treated welded specimen with a semicircular slit at the weld toe (WNS).

Table 4. Peening conditions.

Parameters	Conditions
Air pressure (MPa)	0.5
Radius of needle pin (mm)	1.5
Material of needle pin	High carbon chromium bearing steel (JIS-SUJ2)
Coverage (%)	400

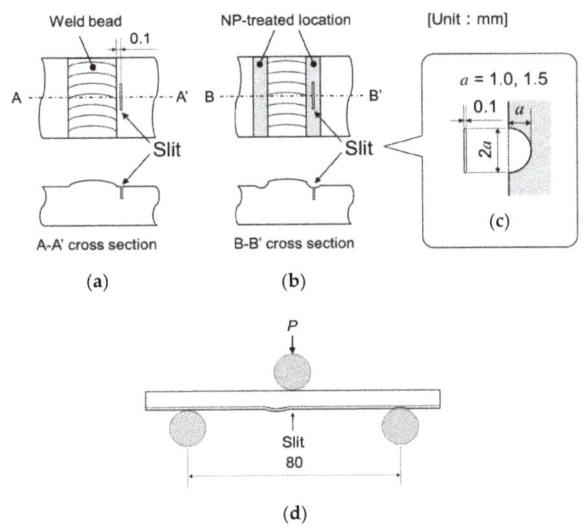

Figure 4. Overview of the slit and loading point: (**a**) position of the slit with respect to the weld bead; (**b**) position of the slit with respect to the weld bead and the location of the NP treatment; (**c**) shape and dimensions of the semi-circular slit where *a* is the slit depth; and (**d**) relationship between the slit position and loading point.

2.3. Fatigue Test Method

Three-point bending fatigue tests were performed with load control using a hydraulic servo testing machine (Shimadzu Corp., Kyoto, Japan). The span length of the three-point bending was 80 mm and the position at which the maximum stress was applied was the same as that of the slit, as depicted in Figure 4d. All tests were performed at a stress of $R = 0.05$ and a frequency of $f = 20$ Hz. The fatigue limit was defined as the maximum stress amplitude at which the specimen could endure 1×10^7 cycles.

2.4. Residual Stress Measurement

The residual stress was measured via the $\cos\alpha$ method using X-ray diffraction from the {211} crystal plane with Cr-Kα radiation. The X-ray residual stress analyzer used was manufactured by Pulstec Industrial Co., Ltd., Hamamatsu, Japan. The diameter of the injected X-ray beam was 0.3 mm and the irradiation range of the X-ray beam was 0.8 mm. In-depth residual stress distributions were obtained by alternately measuring the residual stress on the surface and applying chemical etching to remove the surface layer. Stress correction calculation [18] was performed for each measurement result because stress redistribution occurred during the surface removal.

2.5. Hardness Measurement

In-depth hardness profiles of the weld toe at the surface layer of W and WN were measured using micro Vickers hardness tester manufactured by Shimadzu Corp., Kyoto, Japan. Vickers hardness was measured with an indentation load of 1.961 N and a holding time of 20 s.

2.6. Metal Microstructure Observation

The metal microstructure at the cross-section of W and WN prior to the fatigue test was observed after etching with 3% Nital to clarify the difference between the hardness profile of WN and that of W.

2.7. Finite Element Analysis of the Stress Concentration of the Weld Toe

Finite element (FE) models of W and WN were created based on the shape measurements of the specimens under the fatigue limit using a laser displacement gauge (Panasonic Corp., Kadoma, Japan). Young's modulus and Poisson's ratio were 203 GPa and 0.3, respectively. Elasticity analysis was conducted using universal analysis software (ANSYS ver. 19.0, ANSYS, Inc., Canonsburg, PA, USA).

2.8. Non-propagating Crack Observation

Non-propagating cracks were marked with a heat tint color via heat treatment at 573 K in air. The specimen was compulsorily fractured in a brittle manner using liquid nitrogen after a fatigue crack propagated under excessive cyclic loading.

3. Results and Discussion

3.1. Fatigue Test Results

The fatigue test results (*S-N* diagrams) are presented in Figure 5. The *S-N* diagram plots the nominal stress amplitude (σ_a) against cycles to fracture (N_f). Figure 6 summarizes all fatigue test results as the relationship between the fatigue limit (σ_w) and the depth of the semicircular slit (a). All fatigue test results were shown in Tables A1–A6 in appendix section.

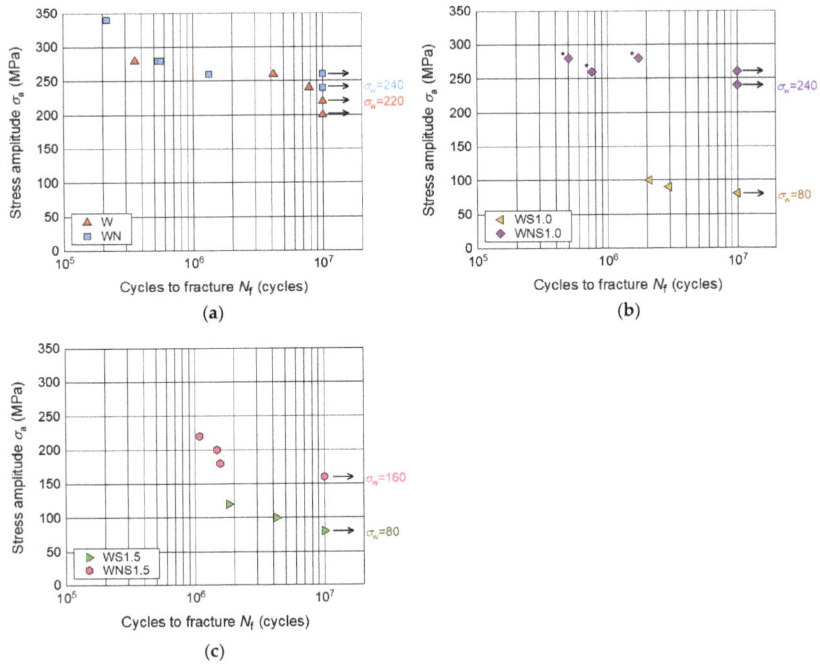

Figure 5. *S-N* diagram of (**a**) W and WN, (**b**) WS1.0 and WNS1.0 and (**c**) WS1.5 and WNS 1.5. The *S-N* diagram plots nominal stress amplitude (σ_a) against cycles to fracture (N_f). Three-point bending fatigue tests were performed with load control. The span length of the three-point bending was 80 mm. All tests were performed at a stress of $R = 0.05$ and a frequency of $f = 20$ Hz. The fatigue test was terminated after 1×10^7 cycles. The asterisks (*) indicate that a fatigue crack occurred at a different location compared to the slit. Fatigue limits σ_w for each type of the specimen were indicated outside the frame of the graph.

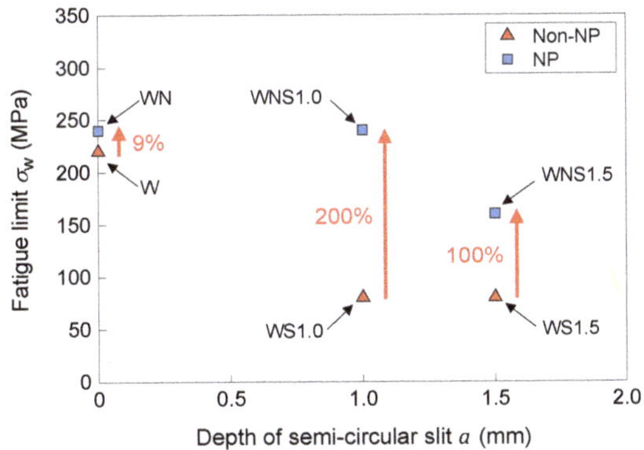

Figure 6. Relationship between the fatigue limit and the semicircular slit depth. The fatigue limit was defined as the maximum stress amplitude at which the specimen could endure 1×10^7 cycles.

In terms of the effect that the semicircular slit had on the fatigue limit, the fatigue limits of WS1.0 and WS1.5 were reduced by 64% compared to that of W. Furthermore, the fatigue limit was significantly lower due to the introduction of a slit, which resulted in a fatigue limit that was less than half that of W. However, the fatigue limit of the specimen treated with NP and WN was 9% higher than that of W. Moreover, the fatigue limits of WNS1.0 and WNS1.5 (i.e., the specimen with a slit treated by NP) were 100% and 200% higher than those of the specimen with a slit but without NP treatment, WS1.0 and WS1.5. The increased fatigue limit of the NP-treated sample with a slit (WNS1.0) was the same as that of the NP-treated sample without the slit (WN). NP remarkably affected the improvement of the fatigue limit for the welded specimen that contained a 1.0-mm-deep slit.

In the present study, a slit is deemed to be harmless if the fatigue test results satisfy either of the following two conditions, which were determined based on a previous study [9,13]:

(a) The fatigue limit of WNS increased to the same level as that of WN.
(b) WNS fractured at a location other than the slit.

The fatigue limit of WNS1.0 was equivalent to that of WN as described above. Therefore, a semicircular slit that is less than $a = 1.0$ mm can be rendered harmless following NP treatment because this situation satisfies condition (a). In addition, a fatigue crack occurred at a different location compared to the slit for all WNS1.0 samples that fractured during the fatigue test, which is indicated by the asterisks (*) in Figure 5b. The fracture surfaces of WNS1.0 and WNS1.5 that were observed using a digital optical microscope are shown in Figure 7. The fracture surface of WNS1.0 is completely different from that of WNS1.5 that fractured from the slit (Figure 7b). Therefore, a semicircular slit of $a = 1.0$ mm was rendered harmless in terms of the crack initiation location.

These results indicate that the fatigue limits containing surface defects with depths less than 1 mm, which are not detected through NDI (as mentioned in Section 1), are considered to be equal to that of WN. Therefore, the reliability of HTS-welded joints can be significantly improved via NP and the problem regarding the reliability of HTS-welded joints that restricts the industrial utilization of HTS can be solved by performing NP.

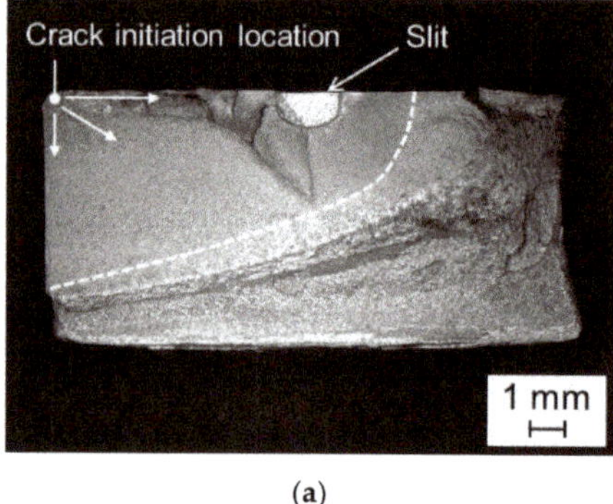

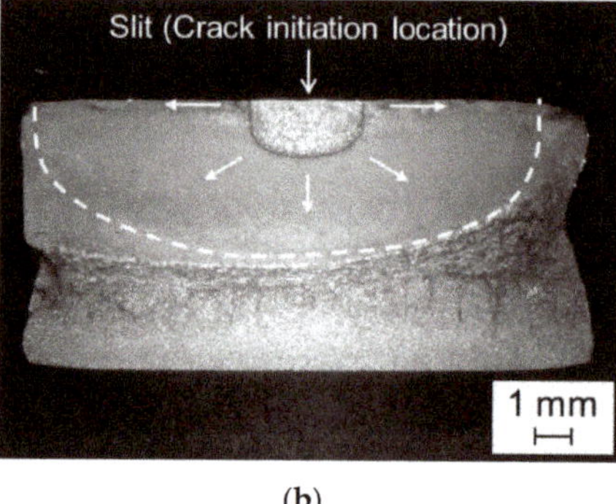

(a) (b)

Figure 7. Fracture surfaces of (**a**) WNS1.0 at $\sigma_a = 260$ MPa and (**b**) WNS1.5 at $\sigma_a = 220$ MPa observed using a digital optical microscope.

3.2. Residual Stress Measurement

The longitudinal residual stress measurement of the weld toe for W and WN was conducted prior to the fatigue test. The residual stress for WN after the fatigue test at a fatigue limit of $\sigma_w = 240$ MPa was also measured to investigate whether the compressive residual stress changed due to cyclic loading.

The residual stress distributions of W and WN are shown in Figure 8. The tensile residual stress was induced at the surface of the weld toe of W due to welding. However, this changed to a large compressive residual stress after NP treatment. The surface compressive residual stress prior to the fatigue test corresponded to approximately 820 MPa. The maximum compressive residual stress corresponded to 850 MPa and was measured at a depth of 0.4 mm. The distance from the surface to the zero residual stress point was approximately 1.3 mm. The residual stress of WN was redistributed due to cyclic loading and the compressive residual stress at the surface was reduced to 520 MPa. Moreover, it was slightly increased at a depth of 0.7–0.9 mm to maintain the equilibrium of the elastic stress field. The maximum compressive residual stress and the zero residual stress point were nearly the same before and after the fatigue test. Section 3.3 explains the reduction of surface compressive residual stress. However, the large and deep compressive residual stress remained high at positions of 1 mm and deeper even after the fatigue test, which greatly contributed to the improvement in the fatigue limit and extension of the acceptable slit size to a depth of 1.0 mm.

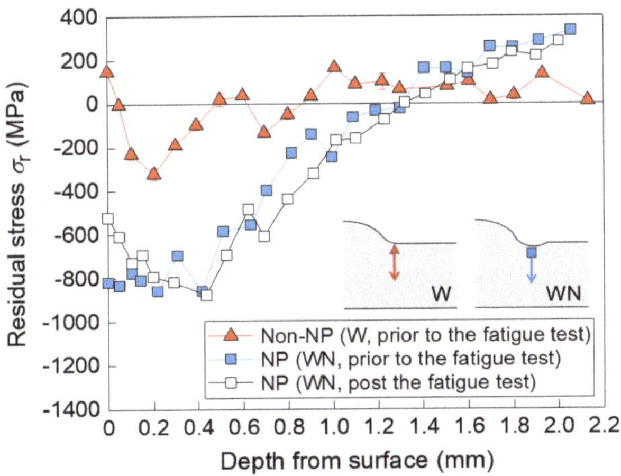

Figure 8. Residual stress distribution at the weld toe. The longitudinal residual stress measurement of the weld toe for W and WN was obtained prior to the fatigue test. The residual stress for WN after the fatigue test at a fatigue limit of σ_w = 240 MPa was also measured.

3.3. Hardness Measurement

In the in-depth hardness profiles of the weld toe at the surface layer of W and WN were measured prior to the fatigue test. The hardness profile of WN after the fatigue test at a fatigue limit of σ_w = 240 MPa was also measured to investigate the effect of cyclic loading on the hardness.

Figure 9 shows the in-depth hardness profiles. The hardness of W slightly changed (mean value of 360 *HV*), but this change was not significant. In contrast, the hardness near the surface of WN prior to the fatigue test corresponded to 510 HV, which was 1.5 times higher than that of W. The hardness of WN gradually decreased with increasing depth until it reached the same hardness of W at that depth, which was approximately 360 HV. Following the fatigue test, the hardness of WN at the surface reduced to 420 HV. At the points that are deeper than 0.15 mm from the surface, the hardness of WN decreased to nearly the same hardness it had prior to the fatigue test. The decrease in the hardness after the fatigue test indicates that the surface layer of WN became easy to yield after the fatigue test because of cyclic softening [19–21] caused by the specimen under a cyclic load. Therefore, the reduction in the surface compressive residual stress mentioned in Section 3.2 was caused by local yielding created by a large compressive residual stress.

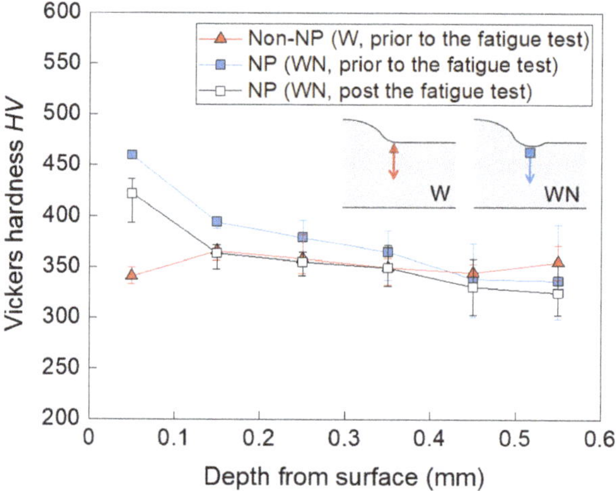

Figure 9. Vickers hardness profiles. In-depth hardness profiles of the weld toe at the surface layer of W and WN prior to the fatigue test were measured. The hardness profile of WN after the fatigue test at a fatigue limit of σ_w = 240 MPa was also measured.

3.4. Metal Microstructure Observation

Figure 10a,b shows the microstructure around the weld toe of W and WN, respectively. The plastic flow was observed at the location of the NP treatment, as shown in Figure 10b. Moreover, the crystal grains of WN (post NP) were micronized compared with the grains of W (without NP). Therefore, the increase in hardness was caused by grain refining and work hardening.

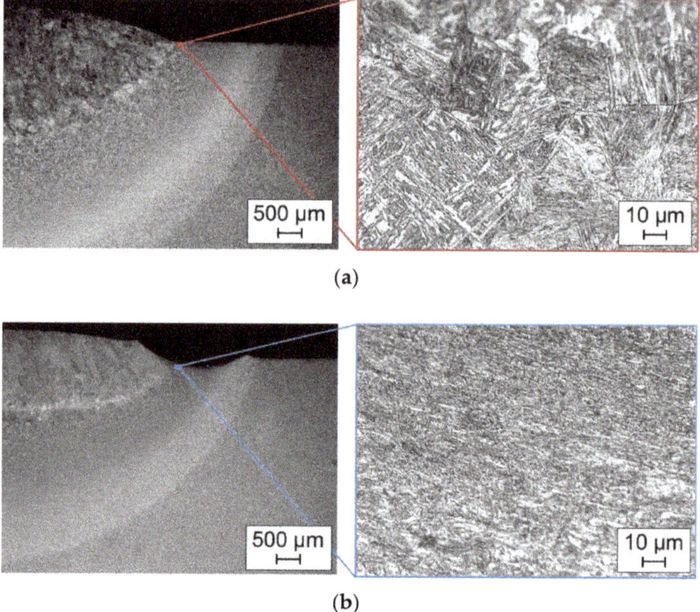

Figure 10. Microscopic image near the weld toe of: (**a**) W; and (**b**) WN.

3.5. Finite Element Analysis of the Stress Concentration of the Weld Toe

Figure 11 shows the FE model of W. The model corresponded to symmetric half models of the welded specimen, which was comprised of a 20-node hexahedral element with a minimum element size of 0.05 mm. Figure 12 shows the elasticity analysis results at a normal bending stress of 200 MPa as the contour of the longitudinal stress around the weld toe. Figure 13 shows the distribution of the longitudinal stress at the weld toe, which was obtained from Figure 12. As shown in Figures 12 and 13, the maximum stress appears to be generated at the bottom of the NP-generated dent of WN and is higher than that of W at the weld toe. The stress concentration factor K_t of the weld toe was calculated to compare the stress with nominal stress. We found that the K_t of W was 1.6 and K_t of WN was 1.8. The reason that the stress concentration increased after NP can be explained by the effect of the dent generated via NP, which is shown in Figure 12b.

Previous studies [14,15] reported that the stress concentration of the weld toe was reduced after NP for welded joints and indicated a high stress concentration factor of K_t of W, which was more than 2.0. However, notably, the stress concentration of the welded joint can increase after NP if a small NP is applied for the welded joint stress concentration at the weld toe.

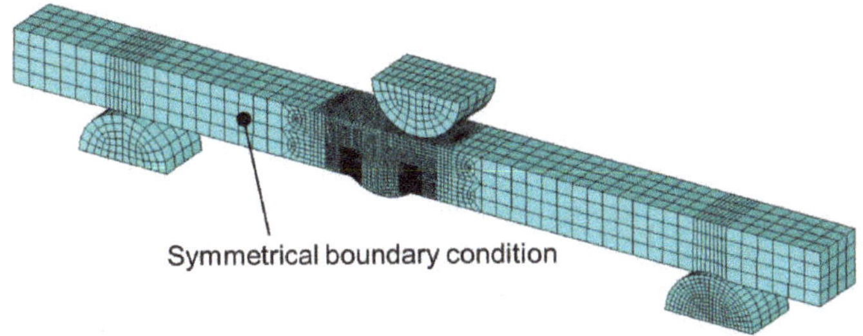

Figure 11. Finite element models of W created based on the shape measurements of the specimens under the fatigue limit using a laser displacement gauge. The model corresponded to symmetric half models of the welded specimen that comprised a 20-node hexahedral element with a minimum element size of 0.05 mm.

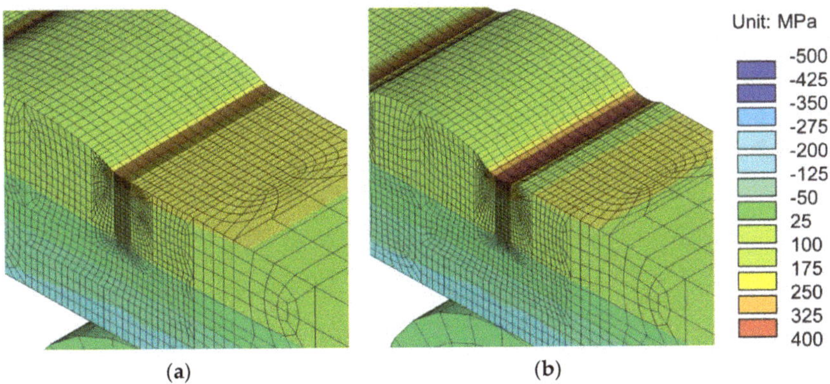

Figure 12. Contour of the longitudinal stress at a normal bending stress of 200 MPa for: (**a**) W and (**b**) WN.

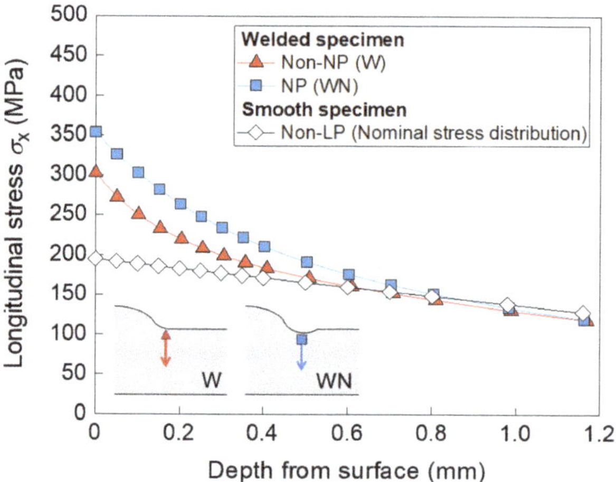

Figure 13. Distribution of the longitudinal stress at the weld toe.

3.6. Dominant Contributing Factor in Fatigue Limit Improvement due to NP

Considering the measurement and analysis results presented in Sections 3.2–3.5, the dominant factor that contributed to the improvement in the fatigue limit and the increase in the acceptable defect depth to less than 1 mm was the introduction of large and deep compressive residual stress during NP. In addition, the increase in hardness at the surface due to work hardening and grain refining caused by NP also contributed to the improvement of the fatigue limit. The negative increase in stress concentration was canceled out by the improvement of residual stress and the increase in hardness.

3.7. Non-propagating Crack Observation

Figure 14 shows the fracture surface around the slit for WNS1.0 that was tested under the fatigue limit. Non-propagating cracks were observed along the slit front, which indicates that the propagation of the fatigue crack initiated from the slit stopped because of large and deep compressive residual stress, as described in Section 3.2.

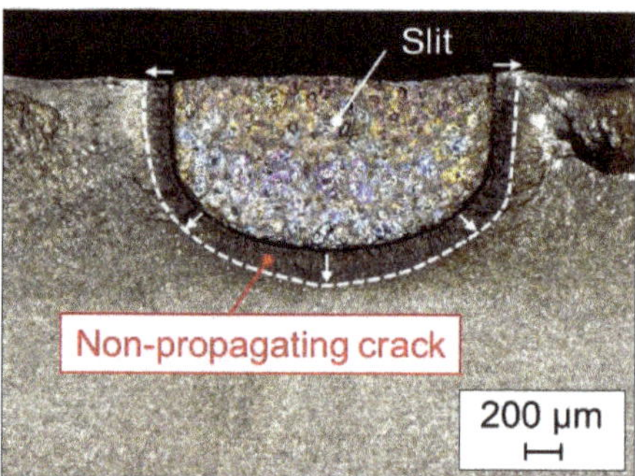

Figure 14. Fracture surface around the slit for WNS1.0 tested under the fatigue limit of $\sigma_w = 240$ MPa. Non-propagating cracks were marked with a heat tint color via heat treatment at 573 K in air. The specimen was compulsorily fractured in a brittle manner using liquid nitrogen after a fatigue crack propagated under excessive cyclic loading.

Therefore, it was clarified that whether the slit affecting the fatigue limit of NP-treated welded joint was determined by the threshold condition for the crack propagation. Additionally, it was found that an acceptable fatigue crack size for the NP-treated welded joint was close to the slit size introduced in the present study, but it was slightly larger than the slit.

4. Conclusions

The present study involved clarifying both the improvement in the fatigue limit and the defect size that can be rendered harmless after NP on a HTS-welded joint containing an artificial semicircular slit on the weld toe. The important knowledge gained through our present study is presented as follows:

(1) The fatigue limit of the defect-free welded specimen increased by 9% after NP at a stress ratio of $R = 0.05$. The fatigue limit of the welded specimen containing a 1.0-mm-deep semicircular slit was significantly increased (200%) by NP as the slit-containing specimen reached the same fatigue limit as that of the defect-free NP-treated welded specimen. This result indicates that the fatigue limits containing surface defects with depths less than 1 mm, which are not detected through NDI, are considered to be equal to that of the NP-treated welded specimen without a defect. Therefore, the reliability of HTS-welded joints can be significantly improved by NP and the problem regarding the reliability of HTS-welded joints that restricts the industrial utilization of HTS can be solved by performing NP.

(2) The dominant factor that contributed to the improvement in the fatigue limit and increase in the acceptable defect size was the introduction of large and deep compressive residual stress during NP. Furthermore, the increase in hardness at the surface due to work hardening and grain refining caused by NP also contributed to the improvement in the fatigue limit.

(3) It was clarified that whether the slit affecting the fatigue limit of NP-treated welded joint was determined by the threshold condition for the crack propagation. Additionally, it was found that an acceptable fatigue crack size for the NP-treated welded joint was close to the slit size introduced in the present study, but it was slightly larger than the slit.

Author Contributions: Conceptualization, R.F. and K.T.; methodology, R.F. and K.T.; peening treatment, M.H.; investigation, R.F.; finite element analysis, R.F.; data curation, R.F.; writing—original draft preparation, R.F.; writing—review and editing, K.T. and M.H.; visualization, R.F.; supervision, K.T.

Funding: This research was funded by the Grant-in-Aid for JSPS Research Fellow (grant number 17J00191).

Acknowledgments: The authors would like to acknowledge Kotoji Ando from the Yokohama National University, Japan, for his valuable comments.

Appendix A

Tables A1–A6 summarizes all fatigue test results. Three-point bending fatigue tests were performed with load control using a hydraulic servo testing machine. The span length of the three-point bending was 80 mm. All tests were performed at a stress of $R = 0.05$ and a frequency of $f = 20$ Hz. The fatigue test was terminated after 1×10^7 cycles.

Table A1. Fatigue test results for W.

Specimen ID	Stress Amplitude σ_a	Cycles to Failure N_f	Fracture Origin
7	280	353,975	Weld toe
4	260	4,163,922	Weld toe
5	240	7,816,232	Weld toe
3	220	>10,000,000	Run-out
6	200	>10,000,000	Run-out

Table A2. Fatigue test results for WS1.0.

Specimen ID	Stress Amplitude σ_a	Cycles to Failure N_f	Fracture Origin
21	100	2,103,232	Slit
23	90	2,963,700	Slit
16	80	>10,000,000	Run-out

Table A3. Fatigue test results for WS1.5.

Specimen ID	Stress Amplitude σ_a	Cycles to Failure N_f	Fracture Origin
97	120	1,820,548	Slit
99	100	4,228,474	Slit
96	80	>10,000,000	Run-out

Table A4. Fatigue test results for WN.

Specimen ID	Stress Amplitude σ_a	Cycles to Failure N_f	Fracture Origin
36	340	214,915	Bottom of dent
27	280	530,242	Bottom of dent
110	280	557,058	Bottom of dent
37	260	>10,000,000	Run-out
113	260	1,323,142	Bottom of dent
35	240	>10,000,000	Run-out
116	240	>10,000,000	Run-out

Table A5. Fatigue test results for WNS1.0.

Specimen ID	Stress Amplitude σ_a	Cycles to Failure N_f	Fracture Origin
100	280	500,542	Outside of the slit
117	260	756,941	Outside of the slit
114	240	>10,000,000	Run-out

Table A6. Fatigue test results for WNS1.5.

Specimen ID	Stress Amplitude σ_a	Cycles to Failure N_f	Fracture Origin
106	220	1,081,373	Slit
108	200	1,473,980	Slit
112	180	1,566,177	Slit
107	160	>10,000,000	Run-out

References

1. Haagensen, P.J.; Maddox, S.J. IIW Recommendations on Post Weld Improvement of Steel and Aluminium Structures. *IIW Comm. XIII* **2001**, *XIII–1815–00*, 29.
2. NETIS New Technology Information System No.CB-120011-A. Available online: http://www.netis.mlit.go.jp (accessed on 27 January 2019).
3. Marquis, G.B.; Mikkola, E.; Yildirim, H.C.; Barsoum, Z. Fatigue strength improvement of steel structures by high-frequency mechanical impact: Proposed fatigue assessment guidelines. *Weld. World* **2013**, *57*, 803–822. [CrossRef]
4. Berg, J.; Stranghoener, N. Fatigue Strength of Welded Ultra High Strength Steels Improved by High Frequency Hammer Peening. *Procedia Mater. Sci.* **2014**, *3*, 71–76. [CrossRef]
5. Wang, T.; Wang, D.; Huo, L.; Zhang, Y. Discussion on fatigue design of welded joints enhanced by ultrasonic peening treatment (UPT). *Int. J. Fatigue* **2009**, *31*, 644–650. [CrossRef]
6. Zhao, X.; Wang, D.; Huo, L. Analysis of the S-N curves of welded joints enhanced by ultrasonic peening treatment. *Mater. Des.* **2011**, *32*, 88–96. [CrossRef]

7. Nose, T. Ultrasonic Peening Method for Fatigue Strength Improvement. *J. JAPAN Weld. Soc.* **2008**, *77*, 210–213. [CrossRef]

8. Yildirim, H.C.; Marquis, G.B. Fatigue strength improvement factors for high strength steel welded joints treated by high frequency mechanical impact. *Int. J. Fatigue* **2012**, *44*, 168–176. [CrossRef]

9. Takahashi, K.; Amano, T.; Hanaori, K.; Ando, K.; Takahashi, F. Improvement of Fatigue Limit by Shot Peening for High Strength Steel Specimens Containing a Crack-like Surface Defect. *J. Soc. Mater. Sci. Japan* **2009**, *58*, 1030–1036. [CrossRef]

10. Takahashi, K.; Amano, T.; Ando, K.; Takahashi, F. Improvement of fatigue limit by shot peening for high-strength steel containing a crack-like surface defect. *Int. J. Struct. Integr.* **2011**, *2*, 281–292. [CrossRef]

11. Houjou, K.; Takahashi, K.; Ando, K. Improvement of fatigue limit by shot peening for high-tensile strength steel containing a crack in the stress concentration zone. *Int. J. Struct. Integr.* **2013**, *4*, 258–266. [CrossRef]

12. Yasuda, J.; Takahashi, K.; Okada, H. Improvement of fatigue limit by shot peening for high-strength steel containing a crack-like surface defect: Influence of stress ratio. *Int. J. Struct. Integr.* **2014**, *5*, 45–59. [CrossRef]

13. Houjou, K.; Takahashi, K.; Ando, K.; Abe, H. Effect of peening on the fatigue limit of welded structural steel with surface crack and rendering the crack harmless. *Int. J. Struct. Integr.* **2014**, *5*, 279–289. [CrossRef]

14. Fueki, R.; Takahashi, K.; Houjou, K. Fatigue Limit Prediction and Estimation for the Crack Size Rendered Harmless by Peening for Welded Joint Containing a Surface Crack. *Mater. Sci. Appl.* **2015**, *06*, 500–510. [CrossRef]

15. Fueki, R.; Takahashi, K. Prediction of fatigue limit improvement in needle peened welded joints containing crack-like defects. *Int. J. Struct. Integr.* **2018**, *9*, 50–64. [CrossRef]

16. Takahashi, K.; Osedo, H.; Suzuki, T.; Fukuda, S. Fatigue strength improvement of an aluminum alloy with a crack-like surface defect using shot peening and cavitation peening. *Eng. Fract. Mech.* **2018**, *193*, 151–161. [CrossRef]

17. Miki, C.; Fukazawa, M.; Katoh, M.; Ohune, H. Feasibility Study on Detection of Fatigue Crack in Fillet Welded Joint. *J. Japan Soc. Civ. Eng.* **1987**, *1987*, 329–337.

18. Society of Automotive Engineers. *Residual Stress Measurement by X-Ray Diffraction-SAE J784a*; SAE International: Warrendale, PA, USA, 1971.

19. Abel, A.; Muir, H. The effect of cyclic loading on subsequent yielding. *Acta Metall.* **1973**, *21*, 93–97. [CrossRef]

20. Thielen, P.N.; Fine, M.E.; Fournelle, R.A. Cyclic stress strain relations and strain-controlled fatigue of 4140 steel. *Acta Metall.* **1976**, *24*, 1–10. [CrossRef]

21. Yoshida, M.; Asano, T.; Fujishiro, Y.; Shirai, Y. Cyclic Softening of the High-strength Steel HT780 Studied by Positron Annihilation Spectroscopy. *Tetsu-to-Hagane* **2005**, *91*, 403–407. [CrossRef]

Article

A Proposal for the Application of Failure Assessment Diagrams to Subcritical Hydrogen Induced Cracking Propagation Processes

Borja Arroyo Martínez *, José Alberto Álvarez Laso, Federico Gutiérrez-Solana, Alberto Cayón Martínez, Yahoska Julieth Jirón Martínez and Ana Ruht Seco Aparicio

LADICIM (Laboratory of Materials Science and Engineering), University of Cantabria, ETS. Ingenieros de Caminos, Av/Los Castros 44, Santander 39005, Spain; alvareja@unican.es (J.A.Á.L.); gsolana@unican.es (F.G.-S.); acayon@adif.es (A.C.M.); juliethjiron0492@gmail.com (Y.J.J.M.); anaruthseco@hotmail.com (A.R.S.A.)
* Correspondence: arroyob@unican.es; Tel.: +34-942-201837

Received: 30 April 2019; Accepted: 6 June 2019; Published: 10 June 2019

Abstract: In this work, an optimization proposal for a model based on the definition of regions for crack propagation by means of the micromechanical comparison by SEM images and its application to failure assessment diagrams (FADs) is presented. It consists in three approaches. (1) The definition of the crack propagation initiation in the elastic-plastic range. (2) A slight modification of the zones in which the FAD is divided for hydrogen induced cracking (HIC) conditions. (3) The introduction of a simple correction for the definition of the K_r coordinate of the FAD to take into account the fracture toughness reduction caused by an aggressive environment, instead of using a fracture parameter obtained from a test in air. For the experimental work, four medium and high strength steels exposed to a cathodic charge and cathodic protection environments were employed, studying two different loading rates in each case, and testing C(T) samples under slow rates in the environment. The study was completed with a subsequent fractographic analysis by SEM. A good degree of fulfilment was appreciated in both materials and environmental conditions, showing the validity of the predictions supplied by the FAD optimization model proposal, which constitutes an advance in the accuracy of the FAD predictive model.

Keywords: failure assessment diagram (FAD); hydrogen induced cracking (HIC); high strength low alloy steels (HSLA); cathodic protection (CP); cathodic polarization or cathodic charge (CC); subcritical propagation; micromechanisms; crack initiation

---

1. Introduction

The steels used in the power production industry, for instance in offshore and oil and gas facilities, have been evolving continuously over the last few decades. The main advances have taken place in improving their mechanical behavior in order to minimize the weight and optimize the transportation; their fracture toughness, in order to avoid problems caused by brittle fracture when working in aggressive environments and low temperatures; or their durability, in order to increase their resistance to subcritical cracking processes, among others [1].

Medium and high strength steels employed in the aforementioned fields are subjected to stress corrosion cracking (SCC) when working in marine environments involving cathodic protection, or sulfide induced cracking due to bacterial activity [2]. In both cases, hydrogen plays a fundamental role in the cracking mechanisms and therefore these steels should be resistant to hydrogen induced cracking (HIC). In these scenarios, the subcritical cracking mechanisms can be dominated by elastic-plastic controlled critical conditions. Hence, the conventional characterization methods based on linear elastic-plastic mechanisms are not useful.

In the last few decades, new methodologies have been developed in order to characterize the SCC and HIC behavior of steels showing a high resistance to subcritical propagation processes [3,4]. These techniques are based on performing a lab test on pre-cracked specimens in the environment at slow rates, in order to obtain its load-crack opening displacement graph (F-COD), and from it parameters such as the toughness in environment or the subcritical crack growth rate [3]. In previous works carried out during the 2000s and 2010s [5–7], the methodology was completed by its direct application to structural design by means of failure assessment diagrams (FADs), a tool widely used in the evaluation of structural safety with highly satisfactory results. These diagrams provide a bi-parametric evaluation of structural integrity considering the combination of fracture and plastic collapse. The application of FAD to environmental processes is still a novel approach, only a few examples being found in the literature [5–7]. For this reason, the present work proposes improvements to the technique described in [5], which is based on a solid model, in order to simplify and universalize its application to HIC scenarios. The application of this analysis to HIC is a challenge of great interest that is not yet extended to standards.

However, as presented in [5], the possibility of subcritical propagation processes in HIC phenomena should be considered when applying FAD in structural design. For these situations, which do not lead to the fracture or collapse of the structure immediately, the subsequent cracked situations taking place during the propagation can be represented in the FAD diagram as a path from the crack initiation to the final failure of the component assessed [5].

In the aforementioned approach [5], the definition of the K_r coordinate of the FAD is made by using the fracture parameter of the material in air (linear-elastic or elasto-plastic) [8,9], $K_{mat} \approx K_{Ic}$, $K_{mat} \approx K_{Jc}$ or $K_{mat} \approx K_{J0.2}$. However, it is well known that the fracture toughness in aggressive environments such as HIC, K_{IHIC}, is usually lower than its equivalent in air. Hence, a more brittle condition takes place [4,5,10,11], which will lead to higher K_r values in the FAD representation of the subcritical crack path propagation [6].

In this work, a novel approach in order to optimize the application of FAD diagrams to the assessment of environmental assisted cracking propagation is proposed. It consists in introducing a simple correction in the methodology exposed in [5] for the definition of the K_r coordinate of the FAD, allowing the fracture toughness reduction caused by HIC to be taken into account without the need to determine K_{IHIC}. It consists of obtaining an approximated K_{mat} value in environment, $K_{mat\text{-}env}$, directly from the load-displacement (P-COD) curve obtained in the test by applying a proportionality between its maximum load and the one from a fracture toughness test in air on the same material.

In this work a new implementation of a crack propagation model over the FAD is also included, showing the capability and applicability of this tool in HIC processes.

All this is done with the aim of optimizing the FAD based prediction model, proposing an improved one whose validity is verified by testing four medium and high strength steels exposed to cathodic charge (CC) and cathodic protection (CP) environments, studying two different loading rates in each case.

The samples are tested by performing slow rate fracture mechanics tests on C(T) specimens, and the study is completed with a subsequent fractographic analysis by scanning electron microscope (SEM) techniques, in order to check the validity of the proposal from a micromechanical point of view.

2. Background

2.1. Behaviour in HIC Conditions

Fracture properties in HIC processes are commonly characterized by testing fracture mechanics specimens in the aggressive environment under study at a constant load or at a constant loading rate [12–14]. This last technique, constant loading rate, using C(T) specimens tested in a slow rate machine is, probably, the most common, as it always leads to the specimen fracture, avoiding uncertainties. Hydrogen diffusion inside the material is promoted by exposure to the environment, normally during a certain amount of time prior to the mechanical solicitation [10–12] as well as during the test. A subsequent SEM fractographic study is often used to reveal the mechanisms behind the cracking process [10–12].

As explained in Figure 1, by applying the analytical methodology [4], based on the GE-EPRI procedure [3], it is possible to obtain crack size values throughout the test from the experimental P-COD curve, and also to determine the value of the J integral in their elastic and plastic components (and indirectly the stress intensity factor). As shown in Figure 1, it is possible to evaluate the crack propagation rate as a function of the stress intensity factor by using the concept of iso-a curves [3]. The method consists in their superposition to the P-COD curve from the test, and their intersection at the point where the crack length, *a*, has the same value as the iso-a curve crossing, as is explained in refernece [4] in detail.

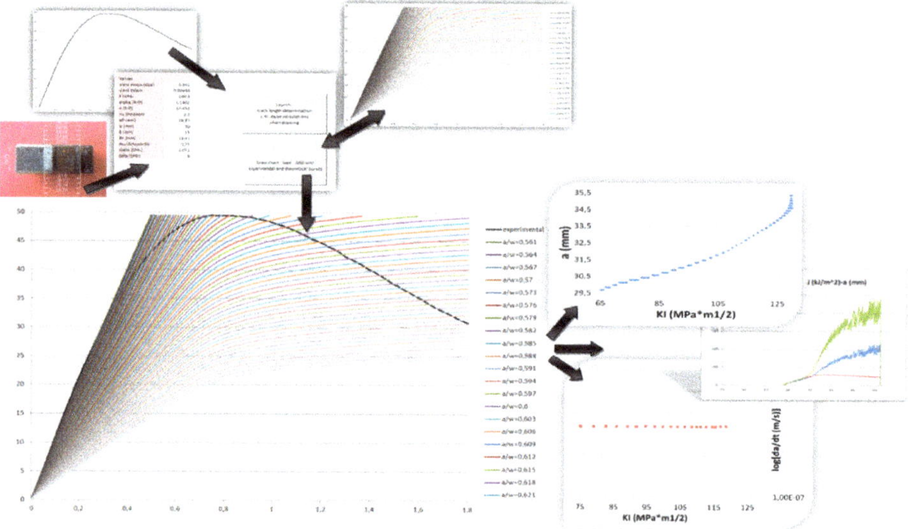

Figure 1. Example of the implementation of GE-EPRI methodology [3,4,15].

2.2. Crack Propagaion Model in HIC

General models attempt to explain the crack propagation modes on HIC conditions of high strength low allow steels (HSLA), mostly intergranular (IG) or transgranular (TG), and the parameters that control them, considering that the presence of hydrogen at the crack tip plays a fundamental role [4,5,16–19]. In [4,18], it is established that the crack propagation in HIC takes place as a series of local isolated fractures, nucleated and developed at the plastic zone ahead of the crack tip. These local fractures nucleate when the obtained plastic strain reaches a critical value determined by the embrittlement produced in this plastic zone by the presence of the hydrogen absorbed by the material, the solubility of the hydrogen being greater in this zone than in the rest of the metallic network. Thus, the model establishes that the crack propagation is controlled by the kinetics of the hydrogen throughout its entrance and diffusion through the crystal network.

The model exposed in [5] proposes that the nucleation of local fractures, which produces propagation, takes place in specific microstructure features or defects occurring either in the grain boundary or inside the grains, constituting traps in which the hydrogen concentration is very high and, consequently, the value of critical strain in them is particularly low.

The type of fracture that arises during propagation, IG or TG, is directly associated to the nucleation process described. If the nucleation occurs at the grain boundary, the fracture will be intergranular (IG), while if it occurs inside the grain it will be transgranular (TG). These different conditions are established through the competence between the grain size, *d*, the position of nucleation, *L**, and the size of plastic zone, r_y (Figure 2). This, finally, leads to limiting expressions for the threshold values of IG or TG propagation as a function of the microstructure (grain size, *d*) and the mechanical behavior of the material (yield stress, σ_y, and Young modulus, *E*). These conditions transferred to macroscopic parameters, such as K_I, lead to the following expressions:

$$IG\ condition \rightarrow KIEAC < 0.85 \cdot \sqrt{\sigma_y \cdot E \cdot d,} \tag{1}$$

$$TG\ condition \rightarrow KIEAC > 0.42 \cdot \sqrt{\sigma_y \cdot E \cdot d.} \tag{2}$$

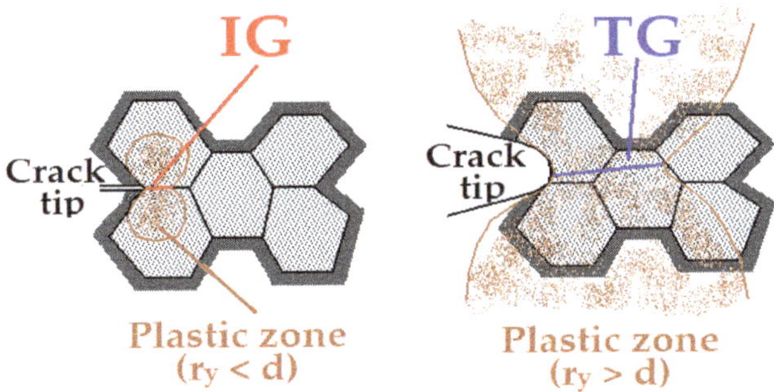

Figure 2. Explanation of intergranular (IG)/transgranular (TG) crack propagation as a function of the grain and the plastic zone sizes.

These conditions can be applied to P-COD representations for a given specimen as a function of crack length, in order to predict the transgranular or intergranular cracking zones. The extensive studies carried out in the area of hydrogen induced cracking [3] provide extensions of the model in order to cover the other micromechanisms observed at the crack propagation of medium and high strength steels. In [5], a new condition is established in order to differentiate the subcritical processes corresponding to low strain situations, as intergranular or pure cleavage. This takes place when it remains associated to local conditions controlled by values of the J integral dominated by its elastic component, J_e, from the cracking paths with tearing or microvoids coalescence associated to values of the J integral in which the plastic component, J_p, is significant.

This condition is defined by the following equation, corresponding in the P-COD representation to a constant displacement value independent of the crack length presented by the specimen [4,20].

$$J_e = 0.95 \cdot J. \tag{3}$$

According to the model from [5], Figure 3 shows the different defined zones in the P-COD plane for steels in HIC. The lines representing the conditions of interganularity (1) and transgranularity (2) define the IG, mixed IG/TG and pure TG (cleavage) zones. The boundary between elastic and plastic domains (3) and the plastic collapse condition for general yielding also define the tearing zone and the zone of plastic instability based on the nucleation growth and coalescence of microvoids.

In Figure 3, a set of P-COD curves that represent tests in one material under different environmental and loading rate conditions are represented. In all the cases, crack propagation is associated to the load decreasing line, which runs, depending on the condition, through different micromechanical zones. The greater the aggressiveness in the testing procedure (higher current density, or voltage or slower displacement rate) the more brittle the conditions achieved.

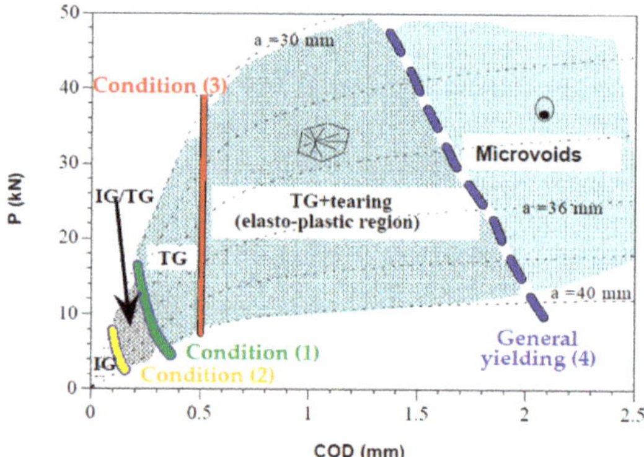

Figure 3. Modeling of the different behavior zones in the load displacement (P-COD) curve [5].

2.3. FAD Implementation of the Model

The approach based on failure assessment diagrams (FADs) establishes that failure in a structure or a component is avoided as long as the structure is not loaded beyond its maximum load capacity defined by using both fracture mechanics criteria and plastic collapse. The classical FAD is a plot of the failure envelope of the cracked component defined in terms of two parameters: K_r and L_r, which are given by the expression below:

$$K_r = \frac{K_I}{K_{mat}},$$ (4)

$$L_r = \frac{P}{P_y} \; or \; L_r = \frac{\sigma_{ref}}{\sigma_y},$$ (5)

where K_r is the ratio of the applied linear elastic stress intensity factor, K_I, and the material fracture toughness, K_{mat}. L_r is the ratio of the applied load, P, and the load that causes plastic yielding of the cracked structure, P_y; an alternative definition in terms of stresses can be given to L_r, being the ratio of the reference stress, σ_{ref}, and the yield stress, σ_y.

The failure envelope is called the failure assessment line (FAL) and is given by expressions proposed in the failure assessment procedures as in [8,20,21] as a function of the material's tensile properties; this line separates the safe and unsafe zones of behavior. Figure 4, obtained by using the software Vindio [22] that incorporates the expressions from [20,21], shows an example of the representation of a FAD diagram.

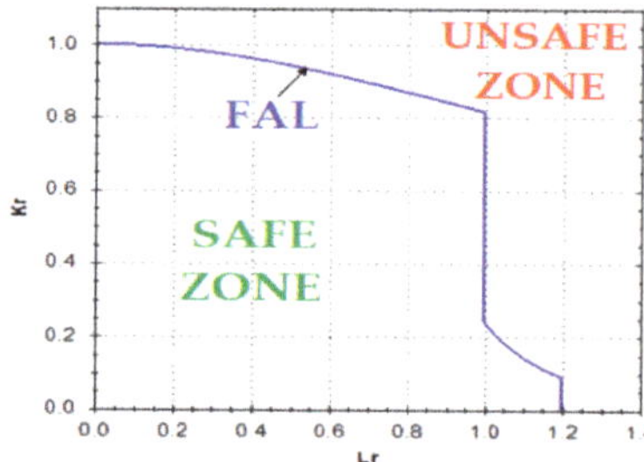

Figure 4. Failure assessment diagram (FAD) graphical representation.

The assessment of a component is based on the relative location of a geometry dependent assessment point respecting the failure line. The FAD diagram analysis allows us to determine the safety conditions without the need to determine if the component operates in small-scale yielding or plastic collapse regime. This is a great benefit, since this distinction would be complicated for a component. The application of this analysis to HIC is a challenge of great interest that is not yet extended to standards.

The previously presented model defines regions with common characteristics in the micromechanisms of cracking in the P-COD diagrams. The transfer from the P-COD representation, with defined zones for the micro-mechanical models, to a FAD diagram constitutes a further step in the unification of the design criteria based on local micro-mechanical approaches, which can be applied in any situation [5].

In [5], it is explained how the use of FAD can be amplified with the inclusion in the representation of the boundary conditions for IG (1) and TG (2) mechanisms in HIC processes, shown as functions of the stress intensity factor, corresponding to straight lines parallel to the horizontal axis (constant K_r) in the FAD representation. The third limit to define the HIC behavior is the initiation of the tearing zone (3), coincident with the development of the plastic component of the J integral. This line in the FAD-diagram corresponds to a representation parallel to the K_r axis (constant L_r). The fourth limit, which defines the general plastification of the residual ligament (4), is coincident with the FAL line at the plastic collapse region, to the right of the failure line. Figure 5 shows the representation of a FAD including the different zones of propagation mechanisms defined by means of the corresponding limit conditions. In all cases, the function of these limits, and therefore the extension of the different zones, depends on the material.

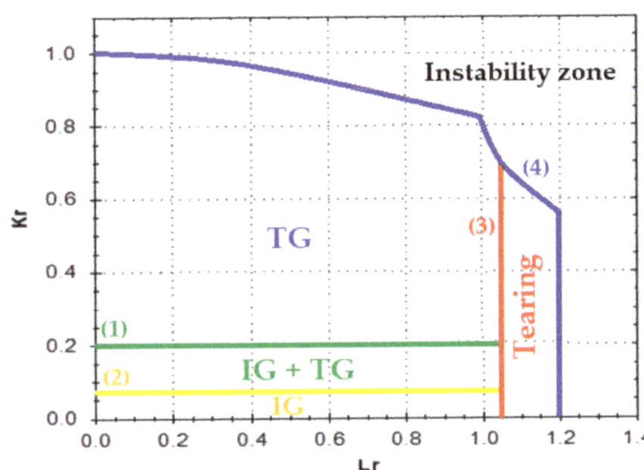

Figure 5. Zones for different hydrogen induced cracking (HIC) mechanisms in a FAD diagram in the original model.

3. Materials and Methods

3.1. Materials

Four materials, from two families, were used in this work:

- Mid-strength low alloy 2.25Cr-1Mo and 3Cr-1Mo-0.25V steels: Designed for working in harsh conditions, commonly used in offshore facilities and water treating reactors. In Figure 6, an image of their bainitic microstructure is presented, showing a relatively small grain size homogeneously distributed.
- High-strength 1Cr-1Ni-1.25Mn and 1Cr-1Ni-0.5Mo steels: Grades R5S and R6, respectively according to [23], obtained by quenching and tempering processes and then forged, employed in the manufacturing of offshore links of chains for mooring lines. In Figure 7, an image of its martensitic and bainitic microstructure is presented, also showing a relatively small grain size homogeneously distributed.

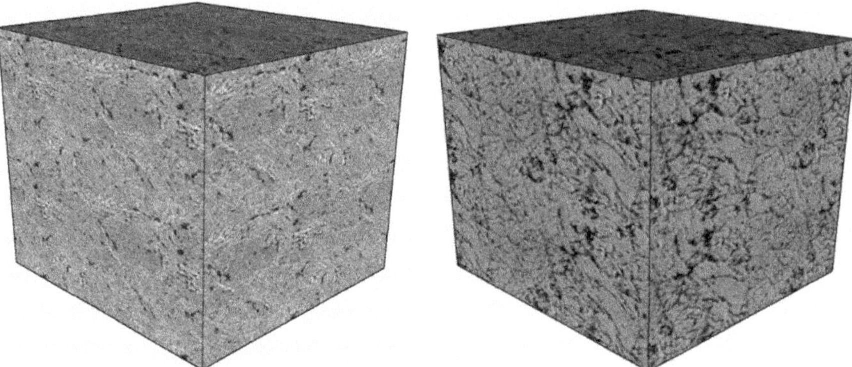

Figure 6. Microstructure of the 2.25Cr-1Mo (**left**) and 3Cr-1Mo-0.25V (**right**).

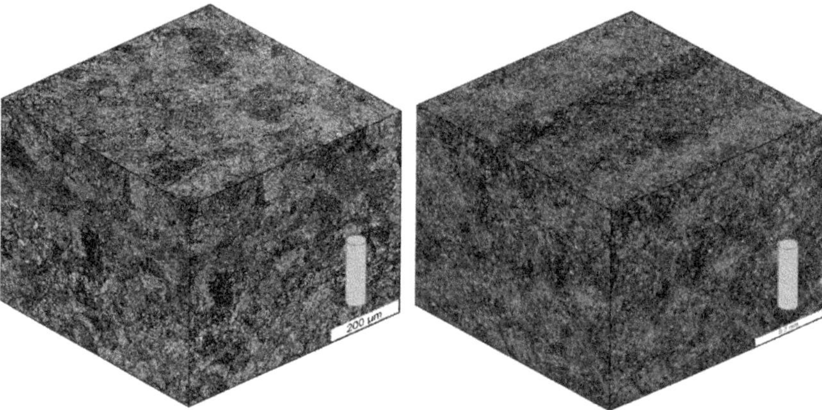

Figure 7. Microstructure of the 1Cr-1Ni-1.25Mn (**left**) and 1Cr-1Ni-0.5Mo (**right**).

In Table 1 the main tensile and fracture properties of all the steels used, as well as their grain size are presented.

Table 1. Main properties of the steels employed.

Parameter	2.25Cr-1Mo	3Cr-1Mo-0.25V	1Cr-1Ni-1.25Mn	1Cr-1Ni-0.5Mo
Sy (MPa)	584	642	1024	1056
Su (MPa)	715	768	1120	1159
E (GPa)	206	211	219	209
ε_{max} (%)	7.3	5.8	6.0	6.2
Ramberg Osgood α	0.592	0.354	0.622	0.485
Ramberg Osgood n	9.4	18.5	17.0	18.0
$J_{0.2}$ (KJ/m^2)	662	241	821	760
$K_{J0.2}$ (MPa·m$^{1/2}$)	387	236	445	418
Grain size (µm)	49	17	48	47

3.2. Environments Applied for Hydrogen Embrittlement

On the one hand, the harsh environments in the oil and gas offshore industry involve the presence of hydrogen, which produces an embrittlement in the exposed materials. On the other hand, in the aforementioned applications, corrosion is always present; protection systems, such as cathodic protection (CP), are applied to minimize its effects, with the handicap that hydrogen embrittlement can also appear.

Hydrogen embrittlement (HE) will always have to be taken into account in these types of industrial facilities. In order to reproduce HE conditions, two different environmental conditions have been employed in this work. The first one, cathodic protection, is usually applied in the accessible parts of the platforms or off-shore structures. The second one, known as cathodic charge or anodic polarization, reproduces local aggressive environments that are impossible to avoid or predict that can seriously affect the structural integrity of the component exposed.

- Cathodic protection (CP) is used to avoid the corrosion phenomena in marine water environments. It involves the use of a sacrificial anode of aluminum (more active than steel), which, in the presence of seawater, is connected to the steel structure. This is the cathode that will be protected from corrosion [24] due to the imposition of a fixed potential between the two, which will maintain the stability of the process. In this study, an aggressive environment of marine water was simulated, consisting of a 3.5% in weight dissolution of NaCl in distilled H_2O [11]. An aluminum anode was employed. The pH was controlled at the range 5.5–5.7 [11] during the whole duration of the tests at room temperature 20–25 °C. The level of cathodic protection (aggressiveness) employed was 1050 mV of fixed potential imposed.

- Cathodic charge (CC), or cathodic polarization, is used to reproduce situations where a huge amount of hydrogen is present, such as acid environments or local situations of hydrogen concentration. It consists in the interconnection, via an acid electrolyte, of a noble material (platinum in this case) and the steel, which will passivate and receive protection due to the fixed current interposed between the two [24]. In this work, the levels of current interposed were of 5 mA/cm^2 (of submerged sample). The aqueous environment employed was an acid electrolyte consisting of a 1N H_2SO_4 solution in distilled H_2O, with 10 mg of an As_2O_3 solution and 10 drops of CS_2 per liter of dissolution. The As_2O_3 solution was prepared following the Pressouyre's method [4,10,25]. The pH was measured in the range 0.65–0.80 during the tests and at room temperature (20–25 °C). The resulting protection causes the hydrogen atoms to be absorbed on the host lattice [26,27], producing local or global embrittlement.

A schema of the cathodic protection (CP) and cathodic polarization (CC) set-ups employed during the tests is shown in Figure 8. In both cases, the aqueous solution is in continuous circulation and/or agitation, as recommended by [12], in order to remove hydrogen bubbles on the specimen surface, avoiding deposits or local environmental conditions.

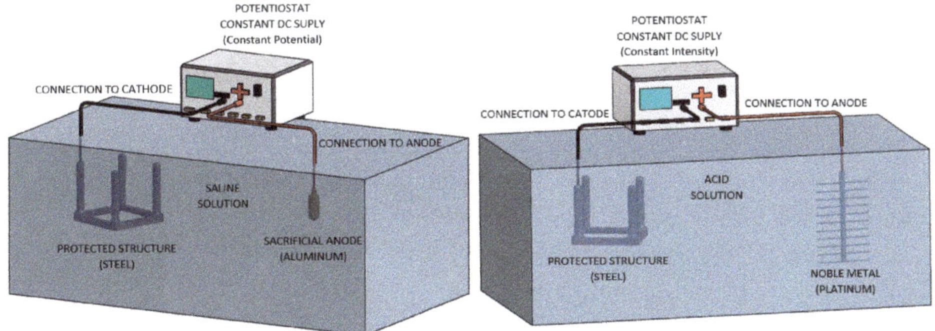

Figure 8. Schema of cathodic protection (left) and cathodic polarization (right) set-ups.

3.3. Optimization Proposals for the Model Presented in Epigraph 2

In the present work, an optimization of the model exposed in [5] (epigraph 2) is presented. It consists of the improvement of the following aspects:

- Slight modification of the zones in which the FAD is divided for HIC conditions. As previously outlined in Section 2 and shown in Figure 5, the FAD diagram is divided into different zones for the different HIC mechanisms taking place (IG, IG/TG, TG, Tearing). On most occasions, depending on the material and the environment, a pure IG zone is very limited; in practice there is a mixed IG/TG mode from the beginning of the propagation. For this reason, in this work, a modification to merge the zones of pure IG and IG + TG into a single one is proposed, modifying the FAD model zones to the one presented in Figure 9 (comparison of Figures 5 and 9).

- Definition of the crack propagation initiation in the elastic-plastic range. For this purpose, the concept of the iso-*a* slope in its straight initial part is employed. Crack initiation is marked at the point in which the iso-a_{ini} curve intersects the P-COD register, the iso-a_{in} being the one that has a slope at the initial straight part equal to 95% of the iso-a_0 (iso-*a* corresponding to the crack length prior to the test, a_0). By using this approach, it is possible to define the crack initiation in an elastic-plastic field, rather than in the linear elastic fracture mechanics (LEFM) one that, for example, ASTM standards [9] recommend. In Figure 10, a graphical representation of the new approach, as well as the LEFM one, is given.

- The last proposed correction is focused on the crack propagation path representation in the FAD. It consists in introducing a modification in the definition of the K_r coordinate ($K_r = K_I/K_{\mathrm{mat}}$) in each of the points that define the crack path propagation in the FAD. It allows the fracture toughness reduction caused by an aggressive environment to be taken into account, without the need to determine K_{IHIC}. The FAD application to environmentally assisted processes leads to predictions far from real behavior, due to the fact that it applies for its calculation a fracture toughness value obtained in non-aggressive scenarios. Applying the aforementioned model, described in Section 2.2, on FAD diagrams and superposing crack path propagations, which is a current trend for predicting propagation and failure in high-plasticity conditions, gave results that did not match with real behavior, as shown in Figure 11. The proposed correction (Figure 11) consists in obtaining an approximate K_{mat} value in environment, $K_{\mathrm{mat\text{-}env}}$, directly from the load-displacement (P-COD) curve obtained in the test. This is done by applying a proportionality between the values of the test in environment and their equivalent values from a fracture toughness test in air, performed on the same material. This correction can be expressed as follows, K_r then being:

$$K_{\mathrm{mat\text{-}env}} = K_{\mathrm{mat}} \cdot \frac{P_{\mathrm{max\text{-}env}}}{P_{\mathrm{max}}}, \tag{6}$$

$$K_r = \frac{K_I}{K_{\mathrm{mat\text{-}env}}}. \tag{7}$$

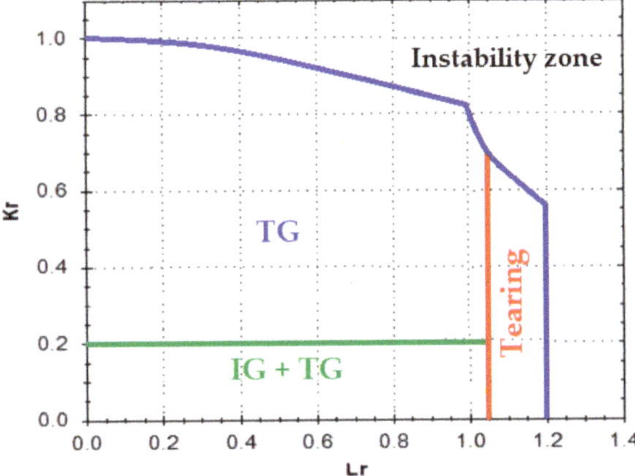

Figure 9. Proposal for zones for different HIC mechanisms in a FAD diagram merging IG with IG + TG.

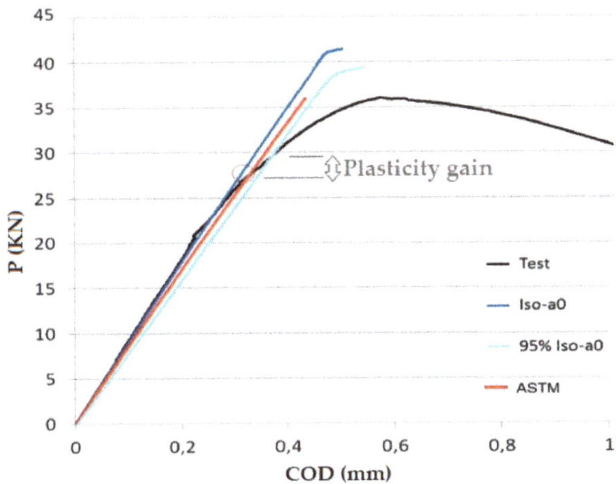

Figure 10. Proposal for the definition of the crack propagation initiation based on the elastic-plastic range.

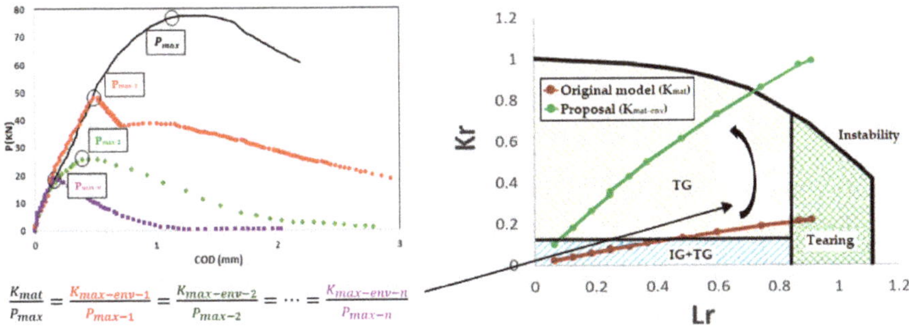

Figure 11. Proposal for the crack propagation crack representation on the FAD.

By applying this modification, it is possible to correct those situations where the embrittlement is substantial. $K_{mat\text{-}env}$ is appreciably lower than K_{mat} so using K_{mat} is on the unsafe side, although it continues to be accurate when the embrittlement is not very great. $K_{mat\text{-}env}$ is close to K_{mat}. As shown in Figure 11, the result from its application will lead to more brittle crack paths in the FAD; L_r values will remain the same while K_r ones will be higher ($K_{\text{mat-env}}$ is lower than K_{mat}), which means that the crack path drawn in the FAD will have a higher slope (K_r/L_r).

3.4. Test Plan

HIC processes were characterized by testing 25 mm thick C(T) specimens according to [9,12,28]. The samples were pre-cracked by $R = 0.1$ fatigue load according to [28], and then 20% side-grooved. After that, the P-COD slow strain tests were performed in the aforementioned environments on the steels selected (epigraph 3.2). In order to reproduce industrial situations, the acid cathodic charge environment (CC) was applied on the mid-strength low alloy 2.25Cr-1Mo and 3Cr-1Mo-0.25V steels, under a current constant intensity, i, of 5 mA/cm^2; the rate effect was studied by testing at displacement rate, V_d, from 10^{-7} to 10^{-9} m/s in each of the possible situations. In order to reproduce off-shore environments, the cathodic protection environment (CP) was also applied on the high-strength 1Cr-1Ni-1.25Mn and 1Cr-1Ni-0.5Mo steels, imposing a constant potential, E, of 1050 mV; in this case rates, V_d, from 6.10^{-8} to 6.10^{-9} m/s were analyzed in each situation. By performing the described combinations of material-environment-rate, a wide range of microstructures and different hydrogen embrittlement scenarios were covered, as shown in Table 2 where the plan tests are presented.

Table 2. Test plan carried out.

Material	Environment (i,E)		Rate (V_d)
2.25Cr-1Mo	Cathodic Charge	5 mA/cm^2	10^{-7} m/s 10^{-9} m/s
3Cr-1Mo-0.25V	Cathodic Charge	5 mA/cm^2	10^{-7} m/s 10^{-9} m/s
1Cr-1Ni-1.25Mn	Cathodic Protection	1050 mV	6×10^{-8} m/s 6×10^{-9} m/s
1Cr-1Ni-0.5Mo	Cathodic Protection	1050 mV	6×10^{-8} m/s 6×10^{-9} m/s

Prior to the test, the samples were exposed to the corresponding liquid aggressive environment for 48 h (1 N H$_2$SO$_4$ solution in H$_2$O under 5 mA/cm^2 for cathodic charge, or simulated marine water under 1050 mV for cathodic protection), in order to polarize them. After that, the mechanical load was applied maintaining the sample exposed to the environment during the whole test. The mechanical testing consisted in the application of a constant loading rate using a slow rate tensile machine of horizontal axis; crack propagation during testing led to the specimen's rupture. In Figure 12, a set-up of the test is presented. After the test, the samples were cleaned using ultrasounds while submerged in acetone and dried. A later SEM fractographic study revealed the mechanism behind the cracking process.

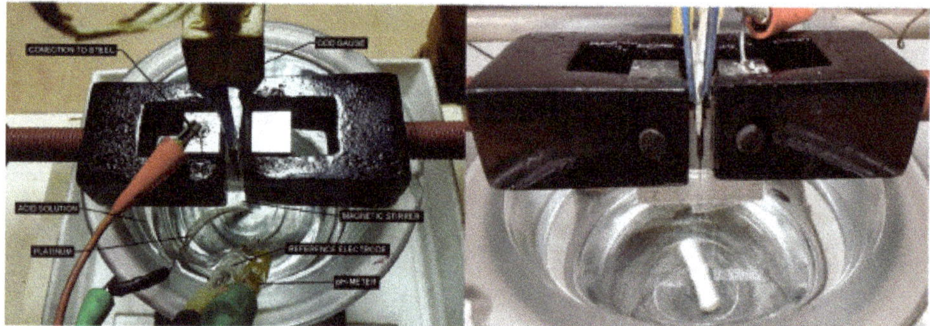

Figure 12. Set-up of the slow rate tests in environment using C(T) specimens.

4. Results

In Figure 13, the P-COD registers from the slow rates tests performed are presented. In Figures 14–17, the subsequent SEM fractographic analysis shows the microstructure during the propagation, positioning the images at their corresponding point of the P-COD register.

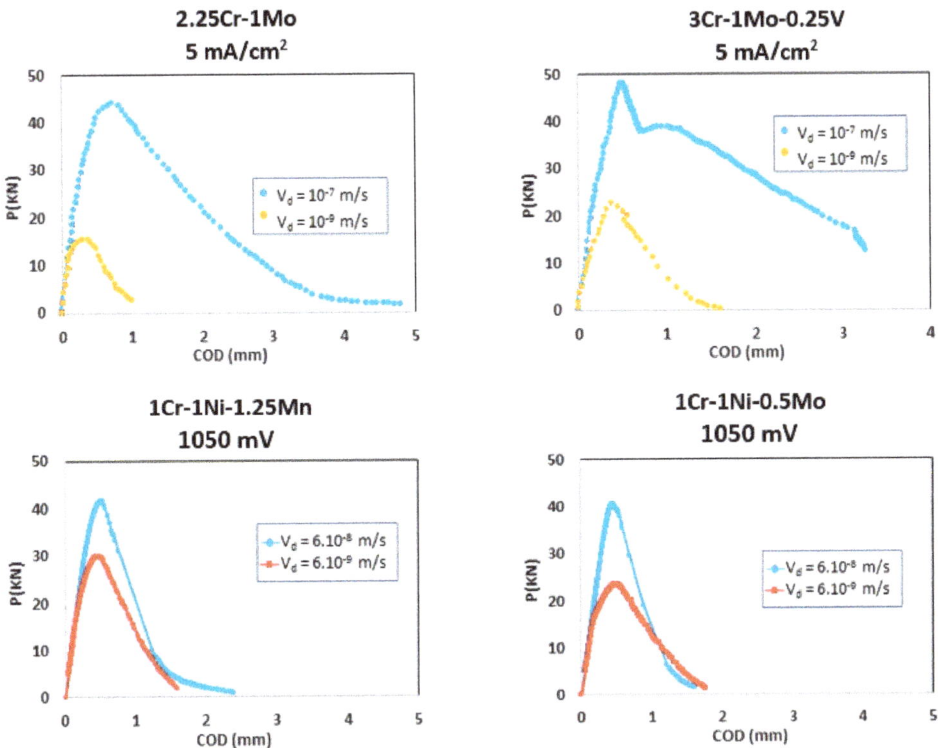

Figure 13. P-COD registers from the slow rate tests in the environment.

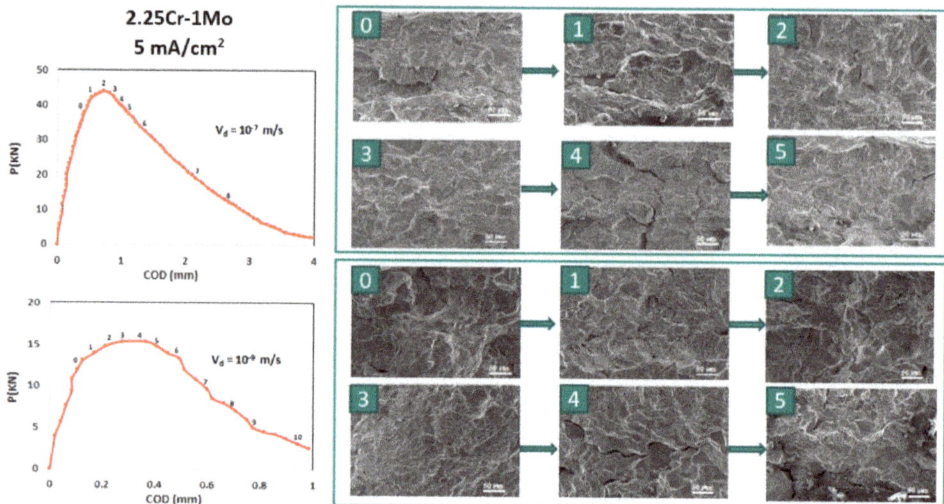

Figure 14. Fractographic images corresponding to different stages of P-COD register from the slow rate tests performed on 2.25Cr-1Mo steel under 5 mA/cm^2 CC environment at 10^{-7} and 10^{-9} m/s.

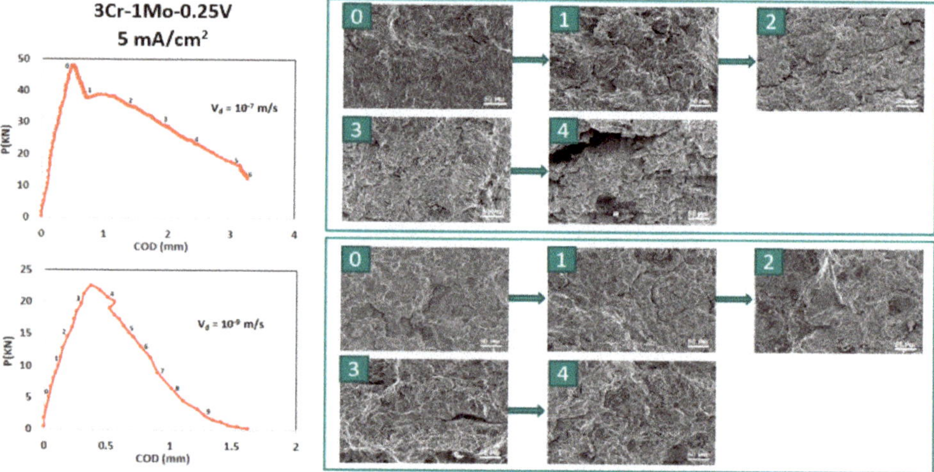

Figure 15. Fractographic images corresponding to different stages of P-COD register from the slow rate tests performed on 3Cr-1Mo-0.25V steel under 5 mA/cm^2 CC environment at 10^{-7} and 10^{-9} m/s.

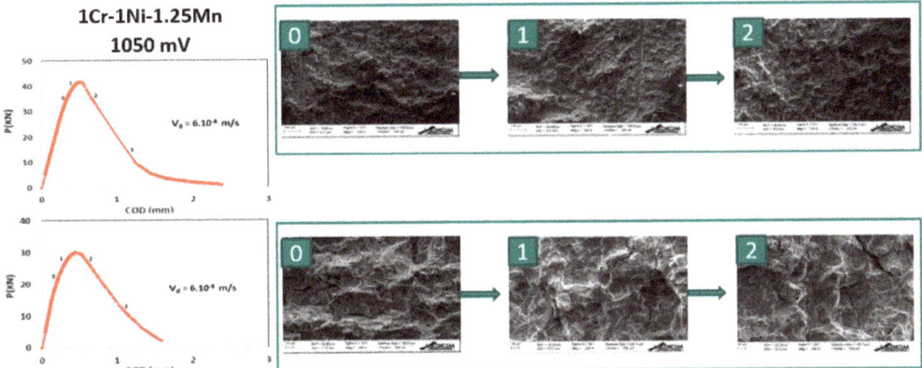

Figure 16. Fractographic images corresponding to different stages of P-COD register from the slow rate tests performed on 1Cr-1Ni-0.25Mn steel under 1050 mV CP environment at 6×10^{-8} and 6×10^{-9} m/s.

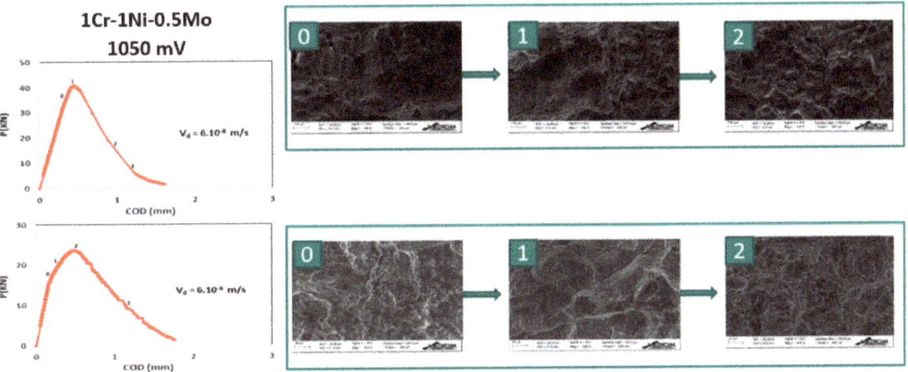

Figure 17. Fractographic images corresponding to different stages of P-COD register from the slow rate tests performed on 1Cr-1Ni-0.5Mo steel under 1050 mV CP environment at 6×10^{-8} and 6×10^{-9} m/s.

From the previous results, in Figures 18–21, the FAD diagrams, including the different zones in which they are for HIC conditions (see Figure 9) and the optimization proposals to the model (epigraph 3.3), are presented. In each case, in addition to the crack propagation path across the different zones, selected SEM micrographs are presented at certain points, in order to be able to verify whether the microstructure taking place along the crack propagation satisfies the proposed modified model.

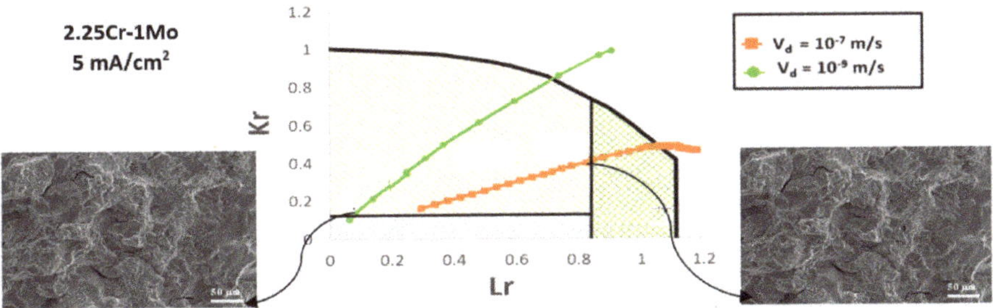

Figure 18. Crack path propagation displayed on the FAD, and fractography, from the slow rate tests performed on 2.25Cr-1Mo steel under 5 mA/cm² CC environment at 10^{-7} and 10^{-9} m/s.

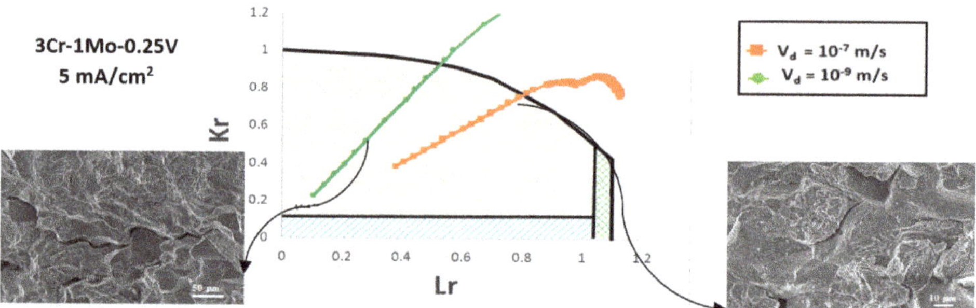

Figure 19. Crack path propagation displayed on the FAD, and fractography, from the slow rate tests performed on 3Cr-1Mo-0.25V steel under 5 mA/cm^2 CC environment at 10^{-7} and 10^{-9} m/s.

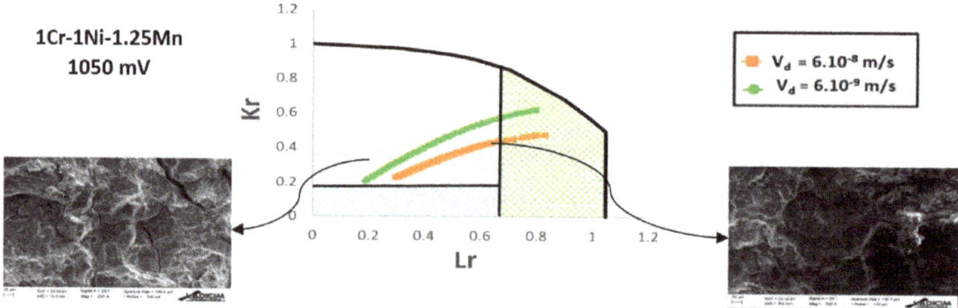

Figure 20. Crack path propagation displayed on the FAD, and fractography, from the slow rate tests performed on 1Cr-1Ni-0.25Mn steel under 1050 mV CP environment at 6×10^{-8} and 6×10^{-9} m/s.

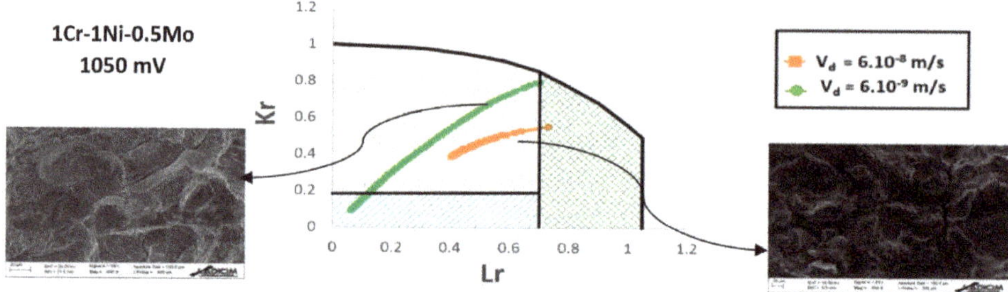

Figure 21. Crack path propagation displayed on the FAD, and fractography, from the slow rate tests performed on 1Cr-1Ni-0.5Mo steel under 1050 mV CP environment at 6×10^{-8} and 6×10^{-9} m/s.

5. Discussion

By examining the P-COD registers, presented in Figure 13, and the SEM fractographic analysis, presented in Figures 14–17, it can be stated that the steels employed in this work have shown susceptibility to the HIC processes used, following the classic rule of dependence on the loading rate. As the rate decreases, the material susceptibility, in a given environmental condition, increases and more brittle micromechanisms are present because the hydrogen available at the crack tip is higher.

By examining Figures 18–21, showing the crack propagation paths across the different zones of the FAD diagrams, it can be stated that:

- 2.25Cr-1Mo steel:

  - 10^{-7} m/s: The initial subcritical process started in the cleavage domain (pure TG, but close to the mixed IG + TG), and the subsequent propagation ran under the cleavage domain (TG) and the plastic domain (Tearing) prior to crossing the failure line (FAL) to the instability zone.
  - 10^{-9} m/s: The process started at the mixed cleavages domain (IG + TG, but very close to the pure TG), and propagated across the TG zone until crossing FAL to the instability zone. No tearing was shown.

- 3Cr-1Mo-0.25V steel:

  - 10^{-7} m/s: The process started at the mixed cleavages domain (IG + TG) and propagated across the TG zone until crossing FAL to the instability zone. No tearing was shown.
  - 10^{-9} m/s: The propagation started at the cleavages domain (IG + TG), and followed across the TG zone until crossing the FAL. No tearing was shown.

- 1Cr-1Ni-1.25Mn steel:

  - 6×10^{-8} m/s: The process started at the TG domain (but close to the mixed IG + TG) and propagated across it up to the plastic domain (Tearing), where the test was stopped (it would have crossed the FAL to the instability zone if continued).
  - 6×10^{-9} m/s: The process started at the TG zone (but very close to the mixed IG + TG) and propagated across the TG zone up to the plastic domain (Tearing), where the test was stopped (it would have crossed the FAL to the instability zone if continued).

- 1Cr-1Ni-0.5Mo steel:

  - 6×10^{-8} m/s: The process started at the TG domain and propagated across it up to the plastic domain (Tearing), where the test was stopped.
  - 6×10^{-9} m/s: The process started at the mixed cleavages domain (IG + TG) and propagated across the TG zone up to the plastic domain (Tearing) very close to the triple point of coincidence of TG-Tearing-FAL (see the green curve on Figure 21), where the test was stopped (it would have crossed the FAL to the instability zone if continued).

In the four combinations of material-environment (two different rates in each case), the rate effect was verified, as previously stated from the analysis of the P-COD curves. In all cases, the rate lowering had an embrittling effect, leading to more brittle path propagation curves in the FAD representation, and thus paths that had a higher slope.

The subcritical propagation grew across the cleavage (TG) region for the most aggressive environments, those with a cathodic charge (CC), until crossing the FAL to the unsafe zone of the FAD. In the 2.25Cr-1Mo and 3Cr-1Mo-0.25V steels under CC, only the least aggressive situation tested (2.25Cr-1Mo at 10^{-7} m/s) invaded the tearing region; these predictions were in accordance with the SEM images obtained.

In the case of the least aggressive environment, cathodic protection (CP), the propagation grew across the cleavage (TG) region towards the tearing region. In the 1Cr-1Ni-1.25Mn and 1Cr-1Ni-0.5Mo steels under CP, only the most aggressive situation tested (1Cr-1Ni-0.5Mo at 6×10^{-9} m/s) invaded the mixed IG + TG region at the crack initiation, and it was also very close to the triple point of coincidence of TG-Tearing-FAL (green curve on Figure 21) in the tearing zone. These predictions were in accordance with the SEM images obtained.

It can be noticed that the extension of the different zones (IG + TG, TG cleavage or tearing) is different for both materials, but the obtained SEM fractographs showed the validity of the predictions supplied by the model. A good degree of fulfilment could be appreciated in both materials and environmental conditions between the cracking processes and predictions, using the optimization proposal described here. Basically, for the materials, environments and rates studied, the model predicts a crack propagation path in the cleavage TG region for the most aggressive environments, up to the FAL line, while for systems less affected by HIC, the crack starts propagating in the mixed IG + TG zone to pass across the cleavage TG zone and develop tearing before crossing FAL.

6. Conclusions and Future Work

In this work, a novel approach in order to optimize the application of FAD diagrams to the assessment of environmental assisted cracking propagation is proposed. In order to do this, firstly, a general review of medium and high strength steel behavior in HIC conditions and the main crack propagation characteristics in these environments was undertaken. Then, the original model for crack propagation regions based on the micromechanical comparison by SEM images and its application to failure assessment diagrams (FAD), presented in [5], has been described in detail, in order to be able to optimize it.

The optimization proposed consisted of three approaches. The first one was the definition of the crack propagation initiation in the elastic-plastic range by using the iso-a slope in its straight initial part. The second one was a slight modification of the zones in which the FAD was divided for HIC conditions, merging pure IG (which can be very limited on some occasions) and the mixed zone of IG + TG into a single region to make it more practical for its application. The third and main approach consisted of introducing a simple correction for the definition of the K_r coordinate of the FAD to take into account the fracture toughness reduction caused by an aggressive environment without the need to determine K_{IHIC}. This consisted of obtaining an approximate K_{mat} value in the environment, $K_{mat-env}$, directly from the load-displacement (P-COD) curve obtained in the test by applying a proportionality between its maximum load and the one taken from a fracture toughness test in air on the same material.

For the experimental work, four medium and high strength steels exposed to cathodic charge (CC) and cathodic protection (CP) environments, studying two different loading rates in each case, have been employed. The samples were tested by performing slow rate fracture mechanics tests on C(T) specimens, and the study was completed with a subsequent fractographic analysis by SEM techniques, in order to verify the validity of the proposals from a micromechanical point of view.

In all the combinations of material-environment-rate, the rate effect was observed; the rate lowering had an embrittling effect, leading to more brittle path propagation curves in the FAD representation. This susceptibility was in compliance with the rules of classic HIC dependence. The obtained SEM fractographs showed the validity of the predictions supplied by the FAD optimization model proposal; a good degree of fulfilment could be appreciated in both materials and environmental conditions.

It was proved that both P-COD and FAD approaches show their capacity to be a useful tool in order to relate the micromechanisms responsible for crack propagation once the different zones have been defined in them. Nevertheless, the FAD, as a tool independent of the component geometry, has proven to be very useful for the assessment of possible HIC processes and their effect on structural integrity. For this reason, the optimization proposed constitutes an advance in the accuracy of the FAD predictive model.

The future work in this field should basically focus on three aspects. First, on the extension of the optimizations proposed to a wider range of materials and environments, and to other specimen geometries (as the FAD is non-dependent of the geometry); second in making their predictions more accurate by incorporating a more precise definition of the K_r coordinate of the crack propagation; finally, as shown in Figure 22 as a proposal for possible future work, a finer micromechanical analysis could help to obtain an improvement in the definition of the correction proposed for K_{mat} ($K_{mat-env}$). Although this involves taking the original model a step further, it consists of a simple proportionally that can be tuned to be more precise.

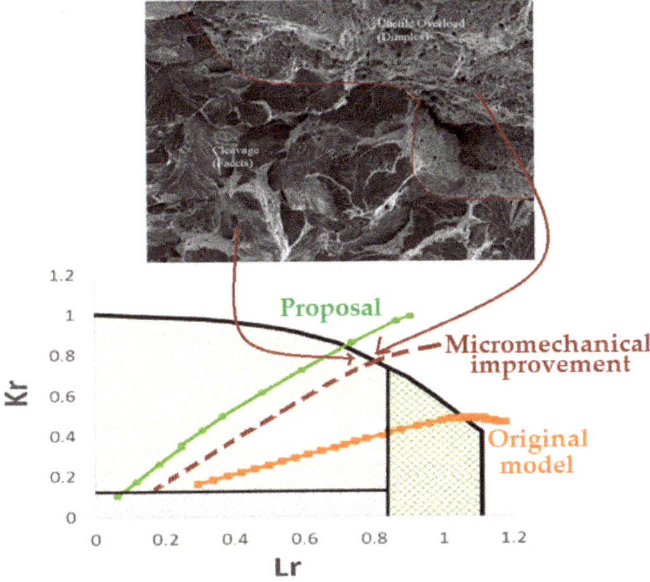

Figure 22. Improvement proposal for possible future work based on the work presented.

Author Contributions: Conceptualization, B.A.M., J.A.Á.L. & F.G.-S.; methodology, B.A.M., J.A.Á.L. & F.G.-S.; validation, B.A.M., A.C.M. & Y.J.J.M.; formal analysis, B.A.M., J.A.Á.L. & F.G.-S.; data curation, B.A.M., A.C.M., Y.J.J.M., & A.R.S.A.; FAD representation, A.R.S.A.; writing—original draft preparation, B.A.M. & J.A.Á.L.; writing—review and editing, B.A.M. & J.A.Á.L.; supervision, J.A.Á.L. & F.G.-S.; project administration, J.A.Á.L. & F.G.-S.; funding acquisition, J.A.Á.L. & F.G.-S.

Funding: This research was funded by CECA, grant number 7210-PE/110, by The Spanish Ministry of Economy and Competitivity, grant number MAT2014-58738-C3-3-R, and by the post-doctoral contracts program of the University of Cantabria, budgetary application 62.0000.64251.

Conflicts of Interest: The authors declare no conflict of interest.

References

1. Navas, J. Diseño y PWHT de aceros avanzados de alta Resistencia para cadenas de líneas de fondeo. Ph.D. Thesis, University of the Basque Country, Bilbao, Spain, 2015.

2. Gabetta, G.; Cole, I. A summary of report for an ESIS working party on fracture control guidelines for environmentally assisted cracking of low alloy steels. *Fatigue Fract. Eng. Mater. Struct.* **1993**, *16*, 603. [CrossRef]

3. Kumar, V.; German, M.; Shih, C.F. *An Engineering Approach to Elastic-Plastic Solid*; Research project EPRI-1931; General Electric Company: Schenectady, NY, USA, 1981.

4. Álvarez, J.A.; Gutiérrez-Solana, F. An elastic-plastic fracture mechanics based methodology to characterize cracking behaviour and its applications to environmental assisted processes. *Nucl. Eng. Des.* **1998**, *188*, 185–202. [CrossRef]

5. Álvarez, J.A.; Gutiérrez-Solana, F. The application of FAD in the assessment of environmental assisted cracking and fracture conditions. In Proceedings of the 13th European Conference on Fractue (ECF 13), San Sebastian, Spain, 8 September 2000.

6. Zhuang, C.; Feng, Y.; Huo, C. Study of the mismatched girth welds of X80 line pipes. In Proceedings of the 2006 International Pipeline Conference, Calgary, AB, Canada, 25–29 September 2006; pp. 187–192.

7. Hasan, S.M.; Khan, F.; Kenny, S. Probabilistic transgranular stress corrosion cracking analysis for oil and gas pipelines. *J. Press. Vessel. Technol. Trans.* **2012**, *134*, 51701. [CrossRef]

8. BS 7910:2013+A1:2015. *Guide to Methods for Assessing the Acceptability of Flaw in Metallic Structures*; BSI: London, UK, 2013.

9. ASTM Standard E-1820-18ae1. *Standard Test Method for Measurement of Fracture Toughness*; ASTM International: West Conshohocken, PA, USA, 2018.

10. Arroyo, B.; Álvarez, J.A.; Lacalle, R.; Uribe, C.; García, T.E.; Rodríguez, C. Analysis of key factors of hydrogen environmental assisted cracking evaluation by small punch test on medium and high strength steels. *Mater. Sci. Eng.* **2017**, *691*, 180–194. [CrossRef]

11. Arroyo, B.; Álvarez, J.A.; Lacalle, R. Study of the Energy for Embrittlement Damage Initiation by SPT Means. Estimation of K_{IEAC} in Aggressive Environments and Rate Considerations. *Theor. Appl. Fract. Mech.* **2016**, *86*, 61–68. [CrossRef]

12. ISO 7539. *Corrosion of Metals and Alloys (Parts 1 to 9)*; ISO: Geneva, Switzerland, different dates for each part.

13. ASTM E-1681. *Standard Test Method for Determining Threshold Stress Intensity Factor for Environment-Assisted Cracking of Metallic Materials*; ASM International: West Conshohocken, PA, USA, 2013.

14. Sedriks, A.J. Stress Corrosion Cracking Test Methods. *Natl. Asoc. Corros. Eng.* **1974**, *543*.

15. Arroyo, B. Caracteización mecáncia de aceros de alta y media resistencia en condiciones de fragilización por hidrógeno mediante ensayos Small Punch. Ph.D. Thesis, University of Cantabria, Santander, Spain, 2017.

16. González, P.; Cicero, S.; Arroyo, B.; Álvarez, J.A. A Theory of Critical Distances based methodology for the analysis of environmentally assisted cracking in steels. *Eng. Fract. Mech.* **2019**, *214*, 134–148. [CrossRef]

17. Thompsonk, A.J.; Bemstein, I.M. *Advances in Corrosion Science and Technology*; Fontana, M.G., Staettle, R.W., Eds.; Springer US: Plenum, NY, USA, 1980; Volume 7.

18. Gutiérrez-Solana, F.; Valiente, A.; González, J.; Varona, J.M. A strain-based fracture model for stress corrosion cracking of low-alloy-steels. *Metall. Mater. Trans.* **1996**, *27A*, 291–304. [CrossRef]

19. Bernstein, I.M.; Thompson, A.W. Effect on metallurgical variables on environmental fracture of steels. *Int. Met. Rev.* **1976**, *21*, 269–290.

20. Álvarez, J.A. Fisuración inducida por hidrógeno en aceros soldables microaleados. Caracterización y modelo de comportamiento. Ph.D. Thesis, University of Cantabria, Santander, Spain, 1999.

21. SINTAP. *Structural Integrity Assesment Procedures for European Industry*; Final Procedure, British Steel Report: Sheffield, UK, 1999.

22. "Vindio 1.0", Software for Structural Integrity Powered by Inesco Ingenieros. Available online: www.inescoingenieros.com (accessed on 6 June 2019).

23. Offshore Standard DNV-OS-E302. *Offshore Mooring Chain*; Det Norske Veritas (DNV): Akershus, Norway, 2008.

24. Hamilton, J.M. The challenges of Deep-Water Artic Development. *Int. J. Offshore Polar Eng.* **2011**, *21*, 241–247.

25. Bernstein, I.M.; Pressouyre, G.M. *Role of Traps in the Microstructural Control of Hydrogen Embrittlement of Steels*; Noyes Publ: Park Ridge, NJ, USA, 1988.

26. Fukai, Y. *The Metal-Hydrogen System: Basic Bulk Properties*; Springer Science & Business Media: Berlin, Germany, 2006.

27. Kirchheim, R.; Pundt, A. Hydrogen in Metals. In *Physical Metallurgy*, 5th ed.; Elsevier: Amsterdam, The Netherlands, 2014; pp. 2597–2705. [CrossRef]

28. ASTM Standard E-399-17. *Standard Test Method for Linear-Elastic Plane-Strain Fracture Toughness KIc of Metallic Materials*; ASTM International: West Conshohocken, PA, USA, 2017.

Article

Baking Effect on Desorption of Diffusible Hydrogen and Hydrogen Embrittlement on Hot-Stamped Boron Martensitic Steel

Hye-Jin Kim [1], Hyeong-Kwon Park [1], Chang-Wook Lee [2], Byung-Gil Yoo [1] and Hyun-Yeong Jung [1,3,*

[1] Research & Development Division, Hyundai-Steel Company, 1480 Bukbusaneopno, Songak-eup, Dangjin-Si, Chungnam 343-711, Korea; khj020911@hyundai-steel.com (H.-J.K.); hkpark@hyundai-steel.com (H.-K.P.); bgyoo@hyundai-steel.com (B.-G.Y.)

[2] Research & Development Division, Hyundai Motor Company, Hwaseong 18280, Korea; lcw8788@hyundai.com

[3] Department of Material Science and Engineering, Korea University, 145 Anam-Ro, Seongbuk-Gu, Seoul 02841, Korea

[*] Correspondence: tamat@hyundai-steel.com; Tel.: +82-41-680-8356

Received: 7 May 2019; Accepted: 29 May 2019; Published: 1 June 2019

Abstract: Recently, hot stamping technology has been increasingly used in automotive structural parts with ultrahigh strength to meet the standards of both high fuel efficiency and crashworthiness. However, one issue of concern regarding these martensitic steels, which are fabricated using a hot stamping procedure, is that the steel is highly vulnerable to hydrogen delayed cracking caused by the diffusible hydrogen flow through the surface reaction of the coating in a furnace atmosphere. One way to make progress in understanding hydrogen delayed fractures is to elucidate an interaction for desorption with diffusible hydrogen behavior. The role of diffusible hydrogen on delayed fractures was studied for different baking times and temperatures in a range of automotive processes for hot-stamped martensitic steel with aluminum- and silicon-coated surfaces. It was clear that the release of diffusible hydrogen is effective at higher temperatures and longer times, making the steel less susceptible to hydrogen delayed fractures. Using thermal desorption spectroscopy, the phenomenon of the hydrogen delayed fracture was attributed to reversible hydrogen in microstructure sites with low trapping energy.

Keywords: hot-press-formed steel; slow strain rate tensile test; thermal desorption spectroscopy; hydrogen-induced delayed fracture

1. Introduction

The global trend of reducing carbon dioxide emissions from vehicles has focused on weight reduction in automobiles. Owing to legislative and customer demand, one significant issue involves weight reduction and crashworthiness properties [1,2]. Recently, hot stamping technology has been increasingly used in automotive structural parts with ultrahigh strength to meet the standards of both high fuel efficiency and crashworthiness [3]. Press-hardened steels are significant owing to their mechanical properties and convenience in fabrication using a hot stamping procedure. A common method of the hot stamping process is direct hot stamping, in which a blank is heated in a furnace, transferred to a press, and subsequently formed and quenched in a cooled die [4]. The full martensitic transformation after manufacturing causes an increase in the extremely high tensile strength. The current strength grade of conventional hot-stamped steel sheets is 1500 MPa, as seen in 22MnB5 steel [3]. Recently, it was reported that 1800-MPa-grade steel was commercialized to produce bumpers in some automotive components, and technical papers were published concerning the development of 2000-MPa-grade hot-stamped steel parts [5,6].

However, a major issue regarding martensitic steels above 1500 MPa that are fabricated using a hot stamping procedure is that they are highly vulnerable to hydrogen delayed cracking caused by the diffusible hydrogen flow through the surface reaction of the coating in a furnace atmosphere [7,8]. The hydrogen delayed fracture is a phenomenon in which a structural part suddenly fails after a time under constant stress. This results from the hydrogen introduced into the steel by the product fabrication process. In the case of the hot stamping process, hydrogen absorption through the coating system occurs during austenitization of the material owing to the presence of moisture in the annealing furnace. In particular, the majority of got formed parts are produced with a hot-dip aluminized coating whose main advantages are scale protection and a corrosion barrier effect. However, it has been reported that the hydrogen uptake must be considered during the hot stamping process with respect to different atmospheric sources on an aluminum coated surface [7].

An important aspect of this problem is the continuous evolution of the material during the hot stamping process. Hydrogen atoms are able to inflow owing to the influence of the dew point, the austenitizing temperature, and the holding time at the austenite temperature in the diffusible hydrogen content [8]. For safe usage, it is essential to improve the resistance by suppressing the hydrogen inflow from the surface and to remove the diffusible hydrogen in the final automobile products. A hydrogen delayed fracture generally involves a certain interaction between the microstructure, hydrogen absorption, and stress level [9–11]. The diffusion behavior of absorbed hydrogen is inhomogeneously distributed in the local microstructure of the material owing to the differences in diffusivity between its complex microstructural characteristics. This plays a role in unique failures resulting from the initiation of microcracks in highly concentrated levels of hydrogen and stress in localized regions such as grain boundaries [12].

Previous studies examined the mechanical strength with respect to the bake hardening effect on hot-pressed 22MnB5 steel used in the hot stamping process [13,14]. This effect is well-known as an improvement in mechanical strength was measured by means of side-crash simulations. The impact bar test was quantified, and the studies focused on aspects of the microstructure. Bake-hardened steels exhibit an increase in yield stress when exposed to temperatures of about 170 °C for 15 or 20 min, such as in the driers of automotive paint lines. This baking hardening process coincides with the tempering and low-temperature heat treatment in martensitic steels [15]. During heat treatment, carbons migrate to dislocations by fixing them. These are the so-called Cottrell atmospheres [16]. This influences the movement of dislocations and requires more effort to cause deformations. However, there is a lack of research on the hydrogen desorption behavior and hydrogen embrittlement properties depending on the baking conditions (including temperature and time parameters) in aluminized hot-press-formed steel with 1800-MPa-grade tensile strength [17].

In this respect, with regard to newly developed hot-stamped martensitic steels with 1800-MPa tensile strength, it is important to understand the diffusible hydrogen behavior for its usage in the automotive industry. To further tailor hot-stamped boron steels with martensitic steels and to ensure their safe usage, it is necessary to investigate the relationship between diffusible hydrogen and delayed fractures depending on the baking conditions and to establish optimum parameters for the baking procedure. In this study, the effect of baking on the desorption of diffusible hydrogen and delayed fracture properties after a hot stamping procedure on aluminum-coated martensitic steel with a tensile strength of 1800 MPa is investigated.

The purpose of this study is as follows: (1) Estimate the diffusible hydrogen inflow during the hot stamping procedure and the desorption behavior of hydrogen depending on the baking conditions, (2) examine the hydrogen embrittlement susceptibility under differential baking conditions after hot stamping, and (3) discuss the relationship between the baking conditions and delayed fractures by observing the fractography characteristics of hot-stamped boron steels. To elucidate the baking effect on the desorption of diffusible hydrogen and the improvement of hydrogen embrittlement in hot-stamped boron martensitic steels, we undertook field emission scanning electron microscopy (FE-SEM) observations combined with energy-dispersive spectroscopy (EDS), transmission electron microscopy (TEM) observations, a slow-strain-rate tensile test (SSRT) after hydrogen charging by controlling the furnace atmosphere during the hot stamping procedure, and a thermal desorption spectroscopy (TDS) analysis.

2. Materials and Methods

The investigated material is an aluminum-silicon-coated cold-rolled steel sheet. The chemical composition of this steel is presented in Table 1. For hot-press forming, Si, Cr, and Mn atoms are added owing to their quenchability, and the B atom plays a role as the strong element that retains the formation of the ferrite phase. Eventually, there is no ferrite when the blank transfers from the furnace to the hot stamping die.

Table 1. Chemical composition of hot-press-formed steel (wt.%).

C	Si	Mn	Cr	B	Ti	Fe
0.300 ± 0.016	0.194 ± 0.028	1.400 ± 0.042	0.200 ± 0.003	0.003 ± 0.0004	0.03 ± 0.0032	Bal.

Cold-rolling to approximately 1.1 mm and a conventional coat protection consisting of 90 wt.% aluminum-10 wt.% silicon prevents scale formation on the surface during the direct hot-stamping process [18]. This coating plays the main role in preventing the formation of severe oxidation on the surface during the hot stamping procedure at temperatures above 950 °C. The microstructures of the coating before and after the hot stamping operations are presented in Figure 1. To transform the microstructure and the mechanical properties, two types of samples were studied: One in the shape of a Japanese Industrial Standards (JIG) 2201 No. 5 test piece used orthogonally in the direction of rolling as a tensile specimen with the milling method for the SSRT test, and the other a square of 300 × 300-mm sheet plates.

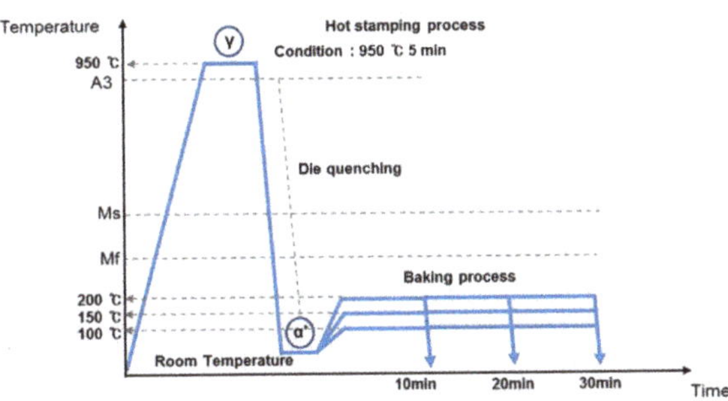

Figure 1. Schematic diagram of hot stamping process and baking conditions.

These samples were austenized for 5 min at 950 °C while keeping the dew point at +5 °C in a furnace. To transform the microstructure and the dramatic hardness value, blanks were austenized for 5 min at 950 °C in a furnace made by Shin-Sung company in Korea. Then, the baking process was performed right after hot stamping. The conditions of baking were as follows: The differential temperatures were 100, 150, and 200 °C (labeled as specimens BH100, BH150, and BH200), and the differential times were 10, 20, and 30 min. Figure 1 shows the hot stamping and baking procedure.

A flat 2 mm × 2 mm specimen was placed in a heating room and heated in an infrared (IR) furnace under vacuum at 20 °C/min to a final temperature of 500 °C. The elevated temperature was recorded by several thermocouples located on the specimen surface. The released hydrogen atoms were detected by quadrupole mass spectroscopy made by PFEIPPER VACUUM company in Germany. The specimens were stored in a liquid nitrogen tank to maintain the charged hydrogen until the hydrogen level was estimated. All specimens were cleaned with ethanol and dried with air. The elevated temperature was recorded by several thermocouples attached to the quartz holder. The released hydrogen gas was carried and monitored by quadrupole mass spectroscopy made by PFEIPPER VACUUM company in Germany in a high-vacuum system. Hydrogen calibration of the TDS facility was performed using a standard-sample NIST 24a titanium alloy before estimating the hydrogen content.

The specimens were stored in a liquid nitrogen tank to retain the charged hydrogen for subsequent level estimation. All specimens were cleaned with ethanol and then air-dried. The hydrogen desorption curves are expressed in ppm/s as a function of the temperature expressed in °C. The reversible hydrogen content was quantified during the heating process by integrating the signal and calculating the accumulated hydrogen content from room temperature to 300 °C, which corresponds to the first peak of the desorption curve. The irreversible hydrogen was quantified from 300 °C to 500 °C, corresponding to the second peak at a higher temperature than the reversible hydrogen desorption range. TDS equipment was used to analyze the hydrogen concentration and elucidate the microstructure relationship by calculating the hydrogen trapping energy [19–21]. The reversible hydrogen content was quantified during the heating process by integrating the signal and calculating the accumulated hydrogen content released from room temperature to 300 °C, which corresponds to the first peak of the desorption curve.

In the case of SSRT, the dramatic variation in the fracture stresses and mechanical properties was quantified in terms of the degree of susceptibility to hydrogen delayed fractures. Based on the JIG 2201 No. 5 standard sample, the elongation in the tensile test was measured at a gauge length of 50 mm without any notching. SSRT for the hydrogen-charged tensile specimens was performed in an air atmosphere at room temperature using a universal tensile testing machine made by Zwick company in Austria [22]. The aluminum intermetallic coating on the surface of the examined samples delayed the diffusion and emission of hydrogen into and out of the specimens during subsequent mechanical tests. Thus, additional injections were not performed with the coating on the surface. The strain rate of the SSRT tests was 0.5 mm/min with a test velocity corresponding to 2.4×10^{-4} /s. The elongation loss, which indicates the degree of reduction in the total elongation after hydrogen charging, was estimated for all test samples depending on the baking conditions. After SSRT, the fracture surface was characterized using the FESEM.

3. Results

3.1. Microstructure and Coating

To analyze the martensitic phases, all specimens were etched in a Nital solution. Micrographs of the hot-stamped steels are shown in Figure 2 before and after the hot stamping procedure. As shown in Figure 2, the as-received initial cold-rolled blanks had a pearlitic-ferritic dual phase before the hot stamping procedure. In the furnace, the microstructure had an austenitic phase. The final automotive parts had a fully martensitic microstructure with a grade of 1800 MPa, and the Al–Si coating was fully transformed into $Fe_x–Al_y–(Si)$ compounds after hot stamping as shown in the Figure 3c,d. Table 2 lists the original mechanical properties with a typical tensile test in a specimen worked from the 300 × 300 sheet plates. The final specimens had a fully martensitic microstructure with a grade of 1800 MPa and an elongation with a value of about 8 El.%.

Table 2. Mechanical properties of base martensitic steel.

Yield Strength (MPa)	Tensile Strength (MPa)	Elongation (El.%)
1302	1875	8.1

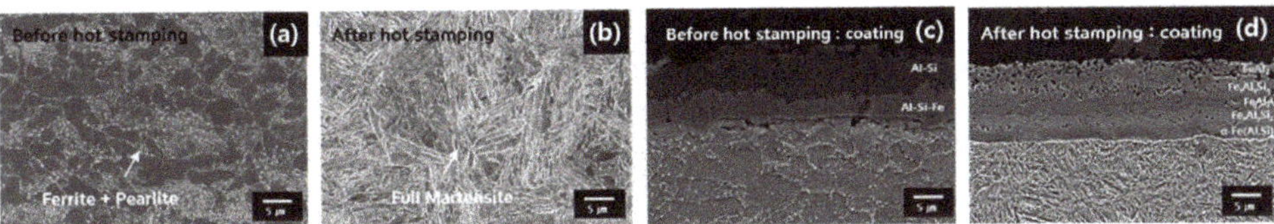

Figure 2. Micrographs of microstructure and coating as hot stamping procedure: (**a**) Microstructure before hot stamping, (**b**) microstructure after hot stamping, (**c**) composition of coating layers before hot stamping and (**d**) composition of coating layers after hot stamping.

3.2. Baking Effect on Hydrogen Embrittlement based on SSRT Results

Right after the hot stamping and baking procedures, SSRTs were conducted for all specimens. Figure 3 shows the stress-strain values of SSRTs for the studied alloys and for differential baking conditions, and plots the actual estimated values of the yield strength, ultimate tensile strength, and elongation corresponding to the stress-strain curves for all specimens. The as-received alloys exhibited elongation reduction (4 El. %) with a hydrogen inflow right after hot stamping without an additional baking process, compared to the above 8%-El. % elongation value, whose specimen had no hydrogen inflow in the steels. Based on the estimated elongation values listed in Table 2, the tendency of elongation loss (El. loss), which indicates the degree of susceptibility of hydrogen embrittlement, was observed in the total elongation as a result of the baking conditions for each specimen.

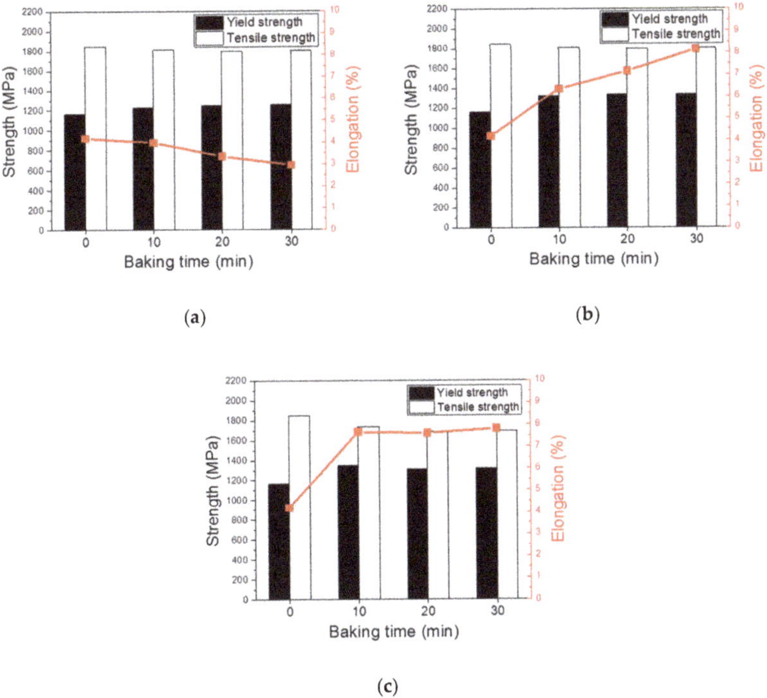

Figure 3. Effect of baking on estimated stress-strain curve depending on temperature and time of baking: (**a**) BH100 conditions, (**b**) BH150 conditions, and (**c**) BH200 conditions.

Figure 3 shows the plots of the estimated elongation including the total elongations of those specimens that were charged with hydrogen in a furnace and those that were removed by the baking process. As shown in Figure 3, the BH100 conditions exhibited a reduction in the elongation loss after hydrogen charging from the identical to the as-received state. However, the BH150 and BH200 conditions exhibited a major increase in the elongation recovery, indicating a decrease in susceptibility to hydrogen embrittlement. When baking at 100 °C, as shown in Figure 3a, there was a slight loss of elongation as the baking time increased from 10 min to 30 min. The tensile strength had similar values for all baking times, and the yield strength increased depending on the baking time.

On the other hand, the BH150 conditions exhibited a proportional increase of reduction in elongation with similar tensile strength as shown in Figure 3b. In the case of the yield strength, the increase in this value occurred after baking for 10 min, but it maintained similar values as the baking time increased. In particular, as shown in Figure 3c, the BH200 conditions showed the most effective impact on the recovery of elongation with a short baking time of 10 min. The figure shows the original elongation in all conditions from 10 min to 30 min for the BH200 conditions. The yield strength increased with the same behavior shown in BH150 specimen, but there was some reduction in the ultimate tensile strength under the target value of 1800 MPa after baking for all specimens.

3.3. Baking Effect on Desorption of Diffusible Hydrogen based on TDS Results

In previous studies, it was reported that hydrogen uptake occurred during the hot stamping process [7,8]. In particular, hot-stamped steels with aluminum coating were considerably more sensitive to hydrogen inflow and hydrogen-induced delayed fractures with significant reductions in elongation and occurrence of plasticity. The significant hydrogen source is mainly affected by the moisture content by the dew point and the gas atmosphere in the furnace. In this study, the reactions of the Al–Si coating to the relationship with the moisture in the air atmosphere and hydrogen uptake in the furnace during the hot stamping process are as follows [7,8,23,24]:

$$2Al + 6H_2O^- \rightarrow 2Al(OH)_3 + 6H, \tag{1}$$

$$2Al + 4H_2O \leftrightarrow 2AlO(OH) + 6H, \tag{2}$$

$$2Al + 4H_2O \rightarrow 2Al_2O_3 + 6H, \tag{3}$$

$$Si + 2H_2O \rightarrow SiO_2 + 6H. \tag{4}$$

To analyze the hydrogen inflow by these reactions in furnace conditions (temperature: 950 °C, dwell time: 5 min, and dew point: +5 °C), the TDS technique was performed right after the hot stamping and baking process. The TDS analysis was generally used to quantify and characterize the concentration and trapping behavior of hydrogen in the material. This consisted of measuring the amount of hydrogen that desorbed from the samples during continuous heating and isothermal holding. To investigate the relationship between the hydrogen solubility behavior, hydrogen concentration, and hydrogen charging time, the hydrogen desorption curves were converted to show the accumulated hydrogen content at each peak for all specimens, as shown in Figure 4. The hydrogen peaks of the desorption curves for all of the alloys occurred mainly at temperatures below 300 °C, indicating that the hydrogen atoms were weakly trapped in reversible trapping sites, such as the lattice, dislocation, and grain boundaries [25].

TDS analysis is often performed to study the behavior of hydrogen trapping in steels. It is possible to estimate the diffusible hydrogen content and to separate the reversible and irreversible hydrogen depending on the temperature range of the desorption. The trap sites in steels are characterized by their binding energy for hydrogen. The diffusible hydrogen corresponding to reversible hydrogen sites with binding energies of <20 kJ/mol is usually considered as that with desorption temperatures. Meanwhile, irreversible hydrogen in immobile trapping sites is defined as that which desorbs from deep traps with binding energies of >60 kJ/mol [25] at high temperatures by acting as irreversible hydrogen trapping sites. For coherent particles, the high-tensile-stress field formation in the matrix surrounding the particle could be more significant in affecting the hydrogen trapping, depending on the level of coherency [26].

In our results, the reversible hydrogen content was mainly decreased by increasing the baking time and temperature, but the irreversible hydrogen content maintained a similar value regardless of the baking conditions for all specimens, as shown in Figure 4. For the BH100 specimens, no considerable difference in the hydrogen content appeared, and there was little desorption of hydrogen as the baking time increased, as shown in Figure 4a. The diffusible hydrogen in the BH150 specimens was released at a proportional tendency as the baking time increased from 10 min to 20 min compared with those under the BH100 conditions and the as-received specimens without baking. However, the desorption behavior between 20 min and 30 min was similar, with the same values of hydrogen content. The desorption behavior of diffusible hydrogen in the BH200 specimens indicated complete removal of the remaining hydrogen after baking regardless of the baking time, and it recovered up to non-charging cold-rolled specimen before hot stamping.

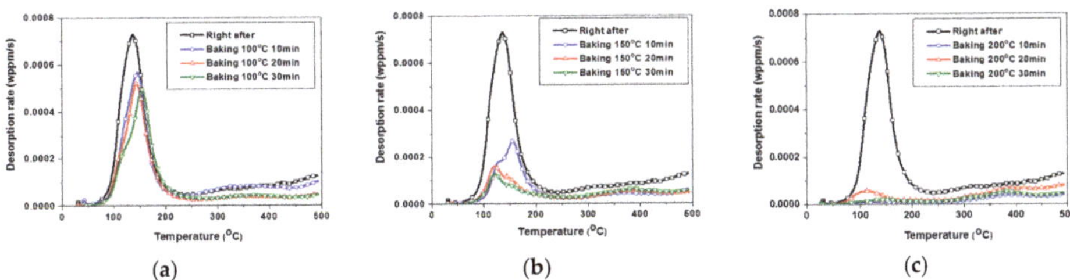

Figure 4. Thermal desorption spectroscopy (TDS) graph of hydrogen desorption behavior corresponding to baking temperature: (**a**) BH100 specimens, (**b**) BH150 specimens, and (**c**) BH200 specimens.

The hydrogen peaks of the desorption curves for all of the alloys occurred mainly at temperatures below 300 °C, acting as diffusible hydrogen with low activation energies. Table 3 and Figure 5 show the accumulated hydrogen content as separated into reversible trapping depending on the hydrogen charging conditions for all specimens below 300 °C. As shown in Figure 5a, the concentration of reversible hydrogen content is decreased by increasing the baking time and temperature compared with that (0.4 wppm) of the as-received state without the baking process. Figure 5b shows the ratio of remaining hydrogen compared with that of the as-received state. In the case of the BH100 conditions, the diffusible hydrogen concentration decreased slightly by up to 0.3 wppm with 80% remaining hydrogen after baking for 30 min. However, based on the SSRT tests, it was difficult for this desorption with a lower reduction of diffusible hydrogen to recover the elongation.

For the BH150 conditions, the diffusible hydrogen was proportionally reduced with an increase in baking time for 20 min and reached about 0.1 wppm, corresponding to a reduction in the remaining hydrogen with a value of about 20%. However, the hydrogen behavior after baking for 30 min was similar to that for 20 min of baking. For the BH200 specimens, the diffusible hydrogen under all baking times was completely removed up to about 0.05 wppm under remaining hydrogen of about 10%, corresponding to the cold-rolled state before hydrogen charging in a furnace. Based on the SSRTs, the recovery of the elongation reduction occurred completely in the BH200 specimens. This demonstrates that a short baking time at 200 °C is most effective for removing the diffusible hydrogen. However, baking at this temperature causes some reduction in the ultimate tensile strength under 1800 MPa, as shown in the above SSRT tests.

Table 3. Concentration of diffusible hydrogen calculated from TDS graphs at 300 °C.

Diffusible Hydrogen Concentration (wppm)			
Specimen	10 min	20 min	30 min
BH100 specimens	0.297	0.285	0.265
BH150 specimens	0.234	0.099	0.088
BH200 specimens	0.032	0.048	0.017

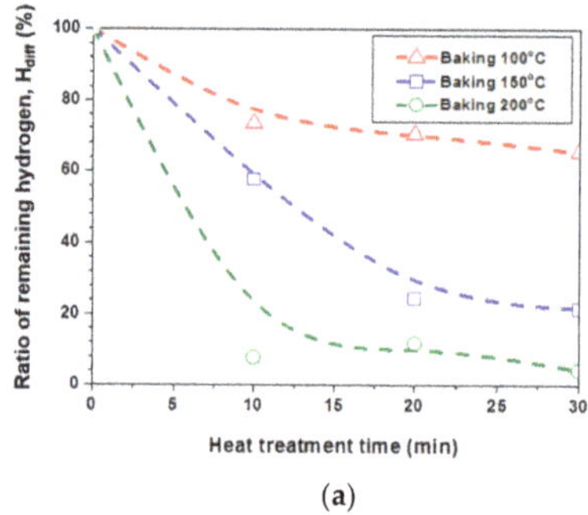

(a)

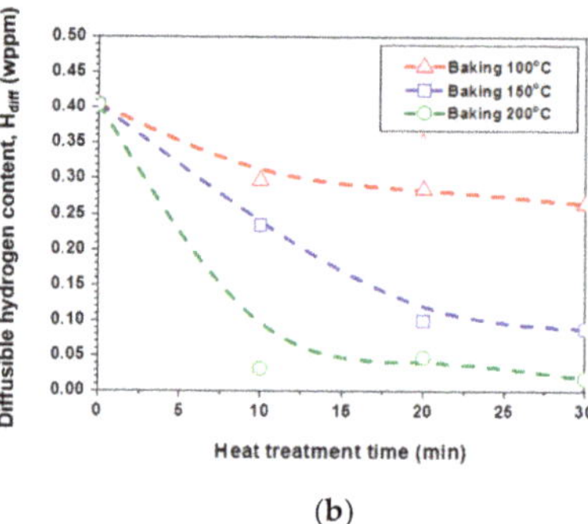

(b)

Figure 5. Diffusible hydrogen behavior on baking conditions: (**a**) Diffusible hydrogen concentration estimated using TDS at 300 °C and (**b**) ratio of remaining diffusible hydrogen.

3.4. Effect of Baking Process and Hydrogen Embrittlement based on Fractography

Fractography is a common technique to elucidate a phenomenon from obtained results [27,28]. As shown in Figures 6–9, the tensile fracture surface at the end of the samples is analyzed to observe the specimen rupture in all specimens. The SEM images that follow in Figures 6–9 are arranged as follows: As-received state without baking process, BH100 specimens, BH150 specimens, and BH200 specimens. Figure 6 shows the as-received specimens, indicating the brittle fracture surface where hydrogen embrittlement plays the main role. This exhibits typically localized brittle fracture modes including mainly cleavage and intergranular fractures from the surface with an obvious difference between the ductile and brittle deformation modes.

It can be observed in Figure 6a that there is a strong tendency between diffusible hydrogen and hydrogen embrittlement as it is all brittle-intergranular showing the severity of degradation. This is significantly related to segregation and pile-up hydrogen concentration at strong reversible trap sites such as the austenite grain boundary, as shown in Figure 6b. This mainly signifies that the crack initiation and propagation procedure in martensitic steels are entirely hydrogen-facilitated. The most pronounced portion of intergranular fractures occurs with a hairline crack in the crack initiation regions.

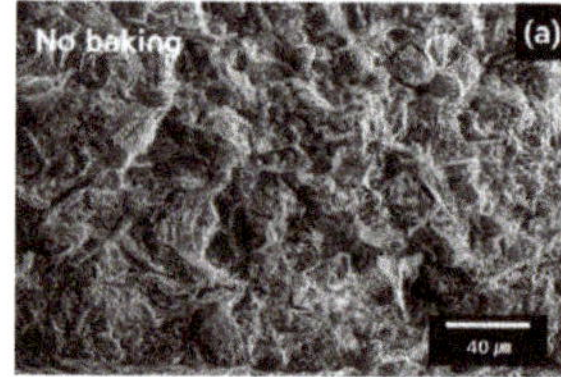

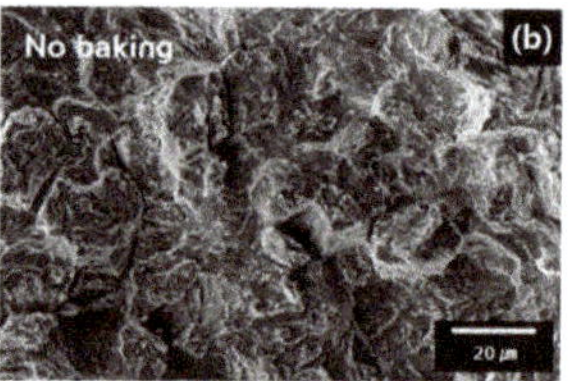

Figure 6. Fractography of slow-strain-rate tensile test (SSRT) specimens after tensile test of as-received state right after hot stamping process: (**a**) Brittle section of fracture surface and (**b**) enlarged section.

Figure 7 shows similar behavior as that of the as-received specimen regardless of baking time in the BH100 specimens, and crack propagation along mainly intergranular fractures with many brittle facets, as shown in Figure 7a,c,e. It seems that there is no strain of grain shape because this sample is cracking under elastic stress. In martensitic steels, the intergranularity combined with quasi-cleavage features are dominant, and flat facets of grains mixed with cleavage are also observed in some regions. Thus, crack initiation and propagation occur under the influence of combined diffusible hydrogen and the stress field in localized regions. As shown in Figure 7b,d,f, the same fracture behavior with intergranular facets and hairline cracks in boundaries is observed for different baking times from 10 to 30 min. This indicates that the BH100 specimens and their conditions are not effective for improving the hydrogen embrittlement in this study.

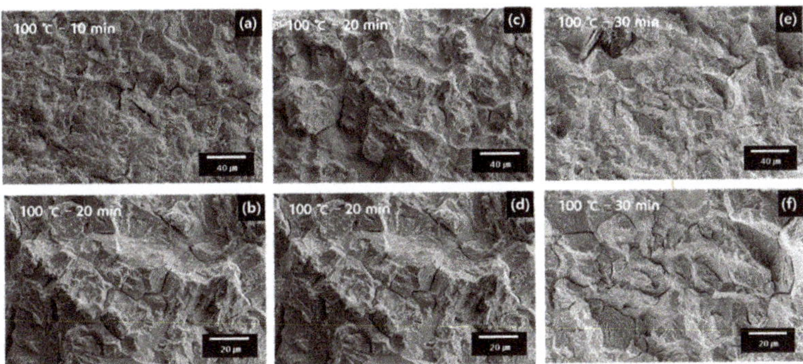

Figure 7. Fractography of SSRT specimens after tensile test of BH100 specimens after hot stamping and baking process: (**a**) Brittle section of fracture surface and (**b**) enlarged section in BH100 at 10 min, (**c**) brittle section of fracture surface and (**d**) enlarged section in BH100 at 20 min, and (**e**) brittle section of fracture surface and (**f**) enlarged section in BH100 at 30 min.

Meanwhile, the BH150 specimens show a threshold baking time in the ductile-to-brittle transition depending on the increase in baking time, as shown in Figure 8. The BH150 specimen at 10 min shows a still-brittle region in the fracture surface, as shown in Figure 8a,b. It appears that the remaining diffusible hydrogen plays a role in the embrittlement in this range of hydrogen concentration. However, in the BH150 specimen at 20 min, a transition of fractography was observed with mainly ductile fractures and microvoid ductile occurrences even though the remaining hydrogen existed at about 20%, as shown in Figure 8c. This indicates a mixed mode of microvoid coalescence, with several large flat regions connected to small areas of small ductile dimples, as shown in Figure 8d.

In addition, the fracture surface of the BH150 specimen at 30 min displays numerous ductile dimples and a small amount of cleavage. This is a similar tendency to the BH150 specimen at 20 min, as shown in Figure 8e,f. Additionally, the crack stabilization region from the center to the edge areas primarily displays microvoid coalescence failure with a ductile dimple. This indicates that the reduction of diffusible hydrogen is effective to generate unusual crack propagation that can stabilize the hydrogen movement and prevent brittle failure during tensile testing by glide plane decohesion, as observed under these conditions.

Figure 8. Fractography of SSRT specimens after tensile test of BH150 specimens after hot stamping and baking process: (**a**) Brittle section of fracture surface and (**b**) enlarged section in BH150 at 10 min, (**c**) brittle section of fracture surface and (**d**) enlarged section in BH150 at 20 min, and (**e**) brittle section of fracture surface and (**f**) enlarged section in BH150 at 30 min.

The BH200 specimens exhibit total recovery with the usual ductile fracture mode at all baking times, as shown in Figure 9. The BH200 specimen in a range of 10 min to 30 min shows no brittle region in any region of the fracture surface, as shown in Figure 9a,c,e. In total conditions of the BH200 specimen, a typical ductile-fracture fine dimple in ultrahigh strength steel without a brittle region in the fracture surface is shown in Figure 9b,d,f. It seems that the removal of diffusible hydrogen under 10% is effective to prevent delayed fractures and plays no role in embrittlement in this range of hydrogen concentration.

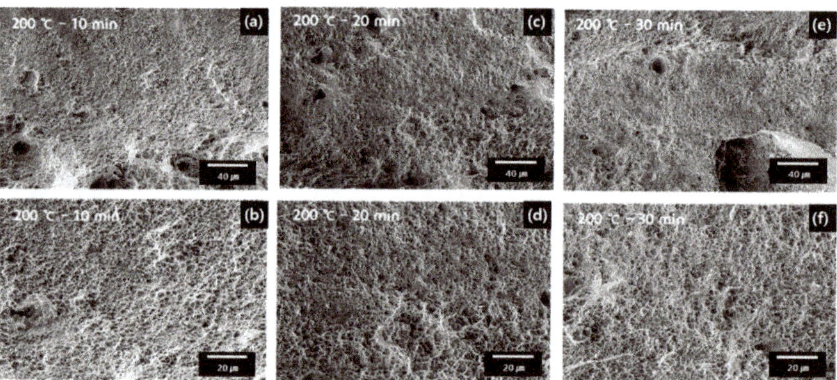

Figure 9. Fractography of SSRT specimens after tensile test of BH200 specimens after hot stamping and baking process: (**a**) Brittle section of fracture surface and (**b**) enlarged section in BH200 at 10 min, (**c**) brittle section of fracture surface and (**d**) enlarged section in BH200 at 20 min, and (**e**) brittle section of fracture surface and (**f**) enlarged section in BH200 at 30 min.

4. Discussion

In previous studies, many effects of baking on microstructures were examined. During the baking process, some precipitations of fine ε-carbides ($Fe_{2.4}C$) consumes solute carbon. Thus, the lattice strains of tetragonal martensite decrease in low-temperature baking processes such as during a tempering effect of 170 °C/20 min [29–31]. Similar results were reported for martensitic C–Mn steels tempered at 200 °C for just 3 min [32]. Since the tensile strength of as-quenched martensite strongly contributes to the amount of carbon, it can be expected that a higher carbon content of steel leads to a greater decrease in the tensile strength during paint baking.

The softening of martensite that occurs during the baking process can be evaluated by comparing the ultimate tensile strength values [33]. Our results show a clear decrease in the R_m values owing to baking under some conditions. This phenomenon can also affect the improvement of hydrogen embrittlement owing to the baking process. However, it is difficult to demonstrate the movement of solute carbon and the reduction of dislocation density in terms of observation of microscopic techniques. Figure 10 shows TEM images of the as-received specimen without baking and the specimen after the baking cycle. It is difficult to observe a meaningful difference depending on the baking as shown in Figure 10b compared with the as-received specimen shown in Figure 10a.

In this study, the hydrogen desorption behavior was examined with respect to its relationship with hydrogen embrittlement rather than from the microscopic aspect. However, it is important to examine the relationship between the two parameters in order to explain the results of the hydrogen desorption behavior. First, the BH100 specimens showed a slight reduction of elongation although the diffusible hydrogen content decreased slightly as the baking time increased. It is assumed that baking at this temperature causes the movement of diffusible hydrogen to defect sites. In addition, there is no decrease in tensile strength at this temperature, which means that the diffusion of carbon solute and release of dislocations rarely occur [34,35]. Based on these results, the BH100 conditions are not effective in improving the hydrogen embrittlement phenomenon.

Second, the BH150 specimens showed a proportional decrease in hydrogen content and a similar reduction in the ultimate tensile strength. The effective remaining hydrogen was 0.1 wppm, and an increase in baking time caused a baking hardening effect in the BH150 specimens. Therefore, in the case of the BH150 conditions, the removal of diffusible hydrogen and the baking hardening effect interact to improve the hydrogen embrittlement property. Third, the BH200 conditions showed complete recovery of elongation regardless of baking time, but the ultimate tensile strength decreased when the baking time increased. The diffusible hydrogen was totally removed in just 10 min. This means that the removal of diffusible hydrogen at this temperature is more effective than the effect of baking hardening.

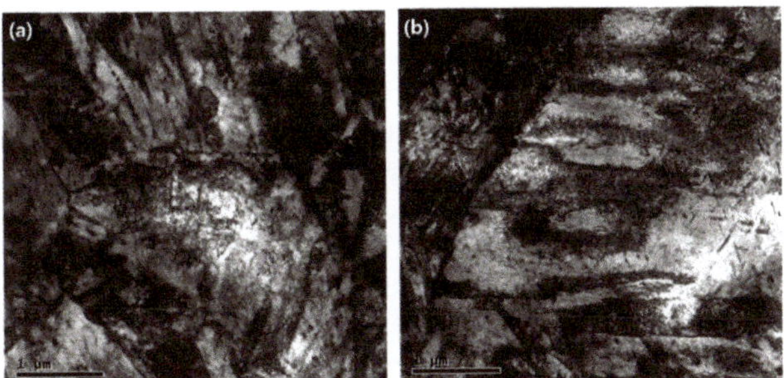

Figure 10. Transmission electron microscopy (TEM) images of martensitic steel: (**a**) As-received specimen after hot stamping and (**b**) specimen after baking process in BH150 for 20 min.

In hydrogen embrittlement, the concentration of dislocations and the localized trapped hydrogen causes fractures along the prior austenite grain boundaries and slip planes of martensitic crystals [36]. It is hypothesized that brittle cracks are the initial source of cracks owing to the high local hydrogen conditions that are hydrogen distribution during deformation. It is also hypothesized that hydrogen accumulation in the necking region during deformation surrounding edge cracks results in mixed quasi-cleavage, and that intergranular fractures are caused by intense shear hand localization, which is reliant upon a local hydrogen concentration high enough to activate the hydrogen-induced localized plasticity mechanism [26,37]. It is mainly assumed that most cases of quasi-cleavage are reliant on a long range of hydrogen diffusion during deformation owing to the need to accumulate hydrogen to activate hydrogen-induced localized plasticity. Purely ductile void coalescence regions were hypothesized to for last, owing to shear overload once the cross section is reduced sufficiently from the surface edge cracks, and the extension of the mixed regions [26].

For these reasons, the relationship between the diffusible hydrogen and dislocation behavior is important in understanding the baking effect and its improvement of hydrogen embrittlement as shown in this study. We can assume that the optimum baking process for recovering the degradation of hydrogen embrittlement and maintaining desired mechanical properties is 150 °C for 20 min, which is the same condition as in paint baking on automobile lines.

5. Conclusions

With regard to newly developed hot stamped martensitic steels with 1800-MPa tensile strength, it is important to understand diffusible hydrogen behavior for its usage in the automotive industry. To further tailor hot-stamped boron steels with a martensitic microstructure and diffusible hydrogen and to ensure their safe usage, it is necessary to investigate the relationship of diffusible hydrogen and delayed fractures depending on the baking condition, and to establish optimum parameters for baking procedures. In this study, the effect of baking on the desorption of diffusible hydrogen and delayed fracture properties after a hot stamping procedure on aluminum-coated martensitic steel with a tensile strength of 1800 MPa was investigated. The following conclusions were drawn with regard to the effect of baking aluminized hot-stamped steel by using hydrogen evaluation and analysis techniques with data obtained from experimental investigations:

(1) The BH100 conditions exhibited a reduction in the elongation loss after hydrogen charging from the identical to the as-received state. However, the BH150 and BH200 conditions exhibited a major increase in the elongation recovery, indicating a decrease in susceptibility to hydrogen embrittlement.

(2) In the BH100 specimens, no considerable difference in the hydrogen content appeared, and there was little desorption of hydrogen as the baking time increased. For the BH150 conditions, the diffusible hydrogen was proportionally reduced as the baking time increased. The desorption behavior of diffusible hydrogen in the BH200 specimens showed complete removal of the remaining hydrogen after baking regardless of baking time, and it recovered to a noncharging cold-rolled specimen before hot stamping.

(3) The softening of martensite that occurred in the baking process can be evaluated by comparing the ultimate tensile strength values. Our results showed a clear decrease in the ultimate tensile strength values owing to baking under some conditions. This phenomenon can also affect the improvement of hydrogen embrittlement owing to the baking process.

(4) The relationship between diffusible hydrogen and dislocation behavior is important in understanding the baking effect and its improvement of hydrogen embrittlement as shown in this study. It is assumed that the optimum baking process for recovering the degradation of hydrogen embrittlement and maintaining the desired mechanical properties occurs at 150 °C for 20 min. This is the same condition as in paint baking on automobile lines.

Author Contributions: H.-J.K., H.-K.P., and C.-W.L., conceived and designed the experiments; H.-J.K. wrote the paper; B.-G.Y. designed materials; H.-J.K. and H.-K.P. performed the experiments; H.-Y.J. proofread English error and revised the manuscript.

Funding: The APC was autonomously funded by Hyundai-Steel company.

Acknowledgments: This work has been studied by the research and development centers of the Hyundai Steel and Hyundai Motor companies in the Republic of Korea.

Conflicts of Interest: The authors declare no conflict of interest. The sponsor had no role in the design of the study; in the collection, analyses, or interpretation of data; in the writing of the manuscript; or in the decision to publish the results.

References

1. Takahashi, M. Sheet steel technology for the last 100 years: Progress in sheet steels in hand with the automobile industry. *ISIJ Int.* **2015**, *55*, 79–88. [CrossRef]
2. Chamisa, A. Development of Ultra High Strength Steels for Reduced Carbon Emissions in Automotive Vehicles. In *Project Outlines*; Chamisa, A., Ed.; University of Sheffield: Sheffield, UK; University of Manchester: Manchester, UK, 2010; p. 11.
3. Karbasian, H.; Tekkaya, A. A review on hot taming. *J. Mater. Process. Technol.* **2010**, *210*, 2103–2118. [CrossRef]
4. Bian, J.; Mohrbacher, H. Novel alloying design for press hardening steels with better crash performance. In Proceedings of the International Symposium on New Developments in Advanced High-Strengh sheet steels, Orlando, FL, USA, 15–18 June 2008; pp. 199–207.
5. Hikida, K.; Nishibata, T.; Kikuchi, H.; Suzuki, T.; Nakayama, N. Development of TS1800MPa Grade Hot Stamping Steel Sheet. *Mater. Jpn.* **2013**, *52*, 68–70. [CrossRef]
6. Tokizawa, A.; Yamamoto, K.; Takemoto, Y.; Senuma, T. Development of 2000 MPa Claass Hot Stamped steel Components with Good Toughness and High Resistance against Delayed Fracture. In Proceedings of the 5th International Conference Hot Sheet Metal Forming of High-Performance Steel, Toronto, ON, Canada, 31 May 2015; pp. 27–35.
7. Georges, C.; Sturel, T.; Drillet, P.; Mataigne, J.M. Absorption/desorption of diffusible hydrogen in aluminized boron steel. *ISIJ Int.* **2013**, *53*, 1295–1304. [CrossRef]
8. Cho, L.; Sulistiyo, D.H.; Seo, E.J.; Jo, K.R.; Kim, S.W.; Oh, J.K.; Cho, Y.R.; De Cooman, B.C. Hydrogen absorption and embrittlement of ultra high strength aluminized press hardening steel. *Mater. Sci. Eng. A* **2018**, *734*, 416–426. [CrossRef]
9. Takai, K.; Watanuki, R. Hydrogen in trapping states innocuous to environmental degradation of high-strength steels. *ISIJ Int.* **2003**, *43*, 520–526. [CrossRef]
10. Takai, K.; Shouda, H.; Suzuki, H.; Nagumo, M. Lattice defects dominating hydrogen-related failure of metals. *Acta Mater.* **2008**, *56*, 5158–5167. [CrossRef]
11. Liu, Q.; Venezuela, J.; Zhang, M.; Zhou, Q.; Atrens, A. Hydrogen trapping in some advanced high strength steels. *Corros. Sci.* **2016**, *111*, 770–785. [CrossRef]
12. Djukic, M.B.; Zeravcic, S.; Bakic, V.G.; Sedmak, A.; Rajicic, B. Hydrogen embrittlement of low carbon structural steel. *Proced. Mater. Sci.* **2014**, *3*, 1167–1172. [CrossRef]
13. Järvinen, H.; Honkanen, M.; Järvenpää, M.; Peura, P. Effect of paint baking treatment on the properties of press hardened boron steels. *J. Mater. Process. Technol.* **2018**, *252*, 90–104. [CrossRef]
14. Castro, M.R.; Monteiro, W.A.; Politano, R. Enhancements on strength of body structure due to bake hardening effect on hot stamping steel. *Int. J. Adv. Manuf. Technol.* **2019**, *100*, 771–782. [CrossRef]
15. Kantereit, H. Baking Hardening Behavior of Advanced High Strength Steels under Manufacturing Conditions. In Proceedings of the SAE Technical Paper, Paris, France, 1 January 2011.
16. Baker, L.J.; Daniel, S.R.; Parker, J.D. Metallurgy and processing of ultra low carbon bake hardening steels. *Mater. Sci. Technol.* **2002**, *18*, 355–368. [CrossRef]
17. Zhang, T.; Hui, W.; Zhao, X.; Wang, C.; Dong, H. Effects of Hot Stamping and Tempering on Hydrogen Embrittlement of a Low-Carbon Boron-Alloyed Steel. *Materials* **2018**, *11*, 2507. [CrossRef]
18. Borsetto, F.; Ghiotti, A.; Bruschi, S. Investigation of the high strength steel Al-Si coating during hot stamping operations. *Key Eng. Mater. Trans. Tech. Publ.* **2009**, *410*, 289–296. [CrossRef]
19. Grabke, H.; Gehrmann, F.; Riecke, E. Hydrogen in microalloyed steels. *Steel Res.* **2001**, *72*, 225–235. [CrossRef]
20. Wei, F.G.; Tsuzaki, K. Quantitative analysis on hydrogen trapping of TiC particles in steel. *Metall. Mater. Trans. A* **2006**, *37A*, 331–353. [CrossRef]
21. Nagao, A.; Martin, M.L.; Dadfarnia, M.; Sofronis, P.; Robertson, I.M. The effect of nanosized (Ti, Mo) C precipitates on hydrogen embrittlement of tempered lath martensitic steel. *Acta Mater.* **2014**, *74*, 244–254. [CrossRef]
22. American Society for Testing and Materials. *Standard Practice for Slow Strain Rate Testing to Evaluate the Susceptibility of Metallic Materials to Environmentally Assisted Cracking*; ASTM G129: West Conshohocken, PA, USA, 2013.

23. Petrovic, J.; Thomas, G. *Reation of Aluminium with Water to Produce Hydrogen*; US Department of Energy: Washington, DC, USA, 2008; pp. 1–26.

24. Bangyikhan, K. Effects of Oxide film, Fe-Rich Phase, Porositiy and their Interactions on Tensile Properties of Cast Al-Si-Mg Alloys. Ph.D. Thesis, Univeristy of Birmingham, Birmingham, UK, 2005.

25. Kim, H.J.; Jeon, S.H.; Yang, W.S.; Yoo, B.G.; Chung, Y.D.; Ha, H.Y.; Chung, H.Y. Effects of titanium content on hydrogen embrittlement susceptibility of hot-stamped boron steels. *J. Alloy. Compd.* **2018**, *735*, 2067–2080. [CrossRef]

26. Nagao, A.; Smith, C.D.; Dadfarnia, M.; Sofronis, P.; Robertson, I.M. The role of hydrogen in hydrogen embrittlement fracture of lath martenisitic steel. *Acta Mater.* **2012**, *60*, 5182–5189. [CrossRef]

27. Hicks, P.D.; Altsteter, C.J. Hydrogen-Enhanced Cracking of superalloys. *Met. Trans.* **1992**, *23A*, 513–522. [CrossRef]

28. Moody, N.R.; Stoltz, R.E.; Perra, M.W. The effect of hydrogen on fracture toughness of the Fe-Ni-Co superalloy IN93. *Met. Trans.* **1987**, *18A*, 1469–1482. [CrossRef]

29. Zhu, X.; Li, W.; Hsu, T.Y.; Zhou, S.; Wang, L.; Jin, X. Improved resistance to hydrogen embrittlement in a high-strength steel by quenching-partitioning-tempering treatment. *Scr. Mater.* **2015**, *97*, 21–24. [CrossRef]

30. Kim, S.J.; Hwang, E.H.; Park, J.S.; Ryu, S.M.; Yun, D.W.; Seong, H.G. Inhibiting hydrogen embrittlement in ultra-strong steels for automotive applications by Ni-alloying. *Mater. Degrad.* **2019**, *3*, 12. [CrossRef]

31. Nakatani, Y.; Higashi, T.; Yamada, K. Effect of tempering treatment on hydrogen-induced cracking in high-strength steel. *Fatigue Fract. Eng. Mater. Struct.* **1999**, *22*, 393–398. [CrossRef]

32. Zhang, Y.J.; Zhou, C.; Hui, W.J.; Dong, H. Effect of C content on hydrogen induced delayed fracture behavior of Mn-B type steels. *J. Iron Steel Res.* **2014**, *26*, 49–55.

33. Kwon, H.J.; Choi, W.S.; Lee, J.; De Cooman, B.C. Bake Hardening Analysis of 22MnB5 PHS by the Impulse Internal-Friction. *CHS2-Series* **2015**, *5*, 499–506.

34. Hirakami, D.; Yamasaki, S.; Tarui, T.; Ushioda, K. Competitive phenomenon of hydrogen trapping and carbon segregation in dislocations introduced by drawing or martensitic transformation of 0.35 mass% and 0.8 mass% C steels. *ISIJ Int.* **2016**, *56*, 359–365. [CrossRef]

35. Sherman, A.M.; Eldis, G.T.; Cohen, M. The aging and tempering of iron-nickel-carbon martensites. *Metall. Trans. A* **1983**, *14*, 995–1005. [CrossRef]

36. Ulmer, D.G.; Altstetter, C.J. Hydrogen-induced strain localization and failure of austenitic stainless steels at high hydrogen concentrations. *Acta Mater.* **1991**, *39*, 1237–1248. [CrossRef]

37. Birnbaum, H.K.; Sofronis, P. Hydrogen-enhanced localized plasticity - a mechanism for hydrogen-related fracture. *Mater. Sci. Eng. A* **1994**, *176*, 191–202. [CrossRef]

Article

Environmentally Assisted Cracking Behavior of S420 and X80 Steels Containing U-notches at Two Different Cathodic Polarization Levels: An Approach from the Theory of Critical Distances

Pablo González *, Sergio Cicero, Borja Arroyo and José Alberto Álvarez

LADICIM (Laboratory of Materials Science and Engineering), University of Cantabria, E.T.S. de Ingenieros de Caminos, Canales y Puertos, Av/Los Castros 44, 39005 Santander, Spain; ciceros@unican.es (S.C.); borja.arroyo@unican.es or arroyob@unican.es (B.A.); jose.alvarez@unican.es (J.A.A.)
* Correspondence: glezpablo@unican.es; Tel.: +34-942-200-928

Received: 29 April 2019; Accepted: 14 May 2019; Published: 16 May 2019

Abstract: This paper analyzes, using the theory of critical distances, the environmentally assisted cracking behavior of two steels (S420 and API X80) subjected to two different aggressive environments. The propagation threshold for environmentally assisted cracking (i.e., the stress intensity factor above which crack propagation initiates) in cracked and notched specimens (K_{IEAC} and K^N_{IEAC}) has been experimentally obtained under different environmental conditions. Cathodic polarization has been employed to generate the aggressive environments, at 1 and 5 mA/cm^2, causing hydrogen embrittlement on the steels. The point method and the line method, both belonging to the theory of critical distances, have been applied to verify their capacity to predict the initiation of crack propagation. The results demonstrate the capacity of the theory of critical distances to predict the crack propagation onset under the different combinations of material and aggressive environments.

Keywords: theory of critical distances; environmentally assisted cracking; hydrogen embrittlement; notch effect; cathodic polarization

---

1. Introduction

It is expected that fossil energies, such as gas and petroleum, will remain the main source of energy for the next two decades. In addition, predictions estimate that energy demand will present an increase of 48% by 2040 [1]. This increasing energy demand has led not only to the development of other energy sources (e.g., nuclear power and renewables), but also to the extraction of fossil energies in more demanding locations. This requires new infrastructure, much of which is operating in increasingly poor conditions (e.g., aggressive environments containing hydrogen sulfides, chlorides and sulfur, among others, elevated temperatures and/or pressures, etc.). The failure of structural components that operate in aggressive environments is often related to environmentally assisted cracking (EAC) processes, such as stress corrosion cracking (SCC) and hydrogen embrittlement (HE) [2,3]. Both phenomena lead to brittle and unexpected failures caused by the degradation of the mechanical properties of the materials [4,5].

In this context, the management of EAC becomes one of the main challenges [6]. The behavior of materials under SCC or HE conditions is a subject of great importance, especially during the material selection process, due to the high cost of the components [7–9]. In addition, repairs and replacements of structural components containing defects involve elevated costs. In this sense, there are numerous situations where the defects jeopardizing the structural integrity of the corresponding component are not cracks, whose radius on the tip tends to zero. This is true in the case of, for example, corrosion defects, which generally have a finite radius (i.e., non-zero) on their tip. If these defects are considered to behave as cracks, the corresponding structural integrity assessments, based on fracture mechanics principles, may be overconservative [10–13]. Hence, it is necessary to develop structural integrity assessment methodologies that take into account the actual behavior of notches, which are here understood as those defects with a finite radius on their tip.

When dealing with the fracture behavior of notches, two different criteria can be distinguished: the global criterion, which is analogous to ordinary fracture mechanics (it compares the notch stress intensity factor with the corresponding notch fracture resistance), and the local criteria, which are based on the stress-strain field around the defect tip. Among the latter, the theory of critical distances (TCD) stands out. The accuracy of the TCD in the prediction of fracture (and fatigue) processes has been widely reported in the literature, especially through its most simple approaches: the point method (PM) and the line method (LM) (e.g., [14–19]).

Recently, the authors have proposed the use of the TCD to analyze EAC processes, providing accurate results [20] in one specific aggressive environment. The aim of this paper is to extend the validation of the use of the TCD in EAC assessments, analyzing the effect of two cathodic current densities on two different steels (S420 and X80) that are commonly used in offshore components, power plants and pipes. The experimental program is composed of 40 C(T) specimens (20 of which are presented for the first time in this paper), with notch radii varying from 0 mm (crack-like defects) up to 2 mm. The aggressive environments have been generated through cathodic charges (or cathodic polarization) at 1 mA/cm^2 (new tests) and 5 mA/cm^2 (tests previously presented in [20]), which are used to cause HE on the steels. The study has been completed with finite element simulations.

2. Theoretical Overview

2.1. The Theory of Critical Distaces

The Theory of Critical Distances (TCD) [14], first presented in the middle of the twentieth century [14,21,22], is a group of methodologies, all of which use a characteristic material length parameter when performing fracture and fatigue assessments. This parameter is called the critical distance, L.

In fracture analysis, the above-mentioned critical distance (L) follows:

$$L = \frac{1}{\pi}\left(\frac{K_{mat}}{\sigma_0}\right)^2 \tag{1}$$

where K_{mat} is the material fracture toughness and σ_0 is a characteristic material strength parameter named the inherent strength, which is usually larger than the ultimate tensile strength and requires calibration. Some critical distance default values for structural steels can be found in the literature [23]. In fatigue analysis, L presents an analogous expression:

$$L = \frac{1}{\pi}\left(\frac{\Delta K_{th}}{\Delta\sigma_0}\right)^2 \tag{2}$$

The two simplest methodologies of the TCD are explained below.

2.1.1. The Point Method (PM)

The PM establishes that fracture occurs when the stress reaches the inherent strength, σ_0, at a distance equal to $L/2$ from the defect tip. The definition of the PM methodology is shown in Figure 1.

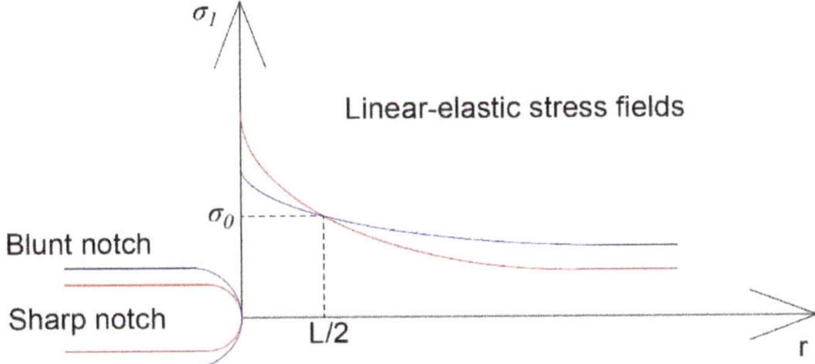

Figure 1. Definition of the point method (PM) methodology. Stress (σ_1)-distance from the notch tip (r) curves due to two different notches in fracture analysis.

In fracture analyses, the mathematical expressions is [14]:

$$\sigma\left(\frac{L}{2}\right) = \sigma_0 \tag{3}$$

whereas in fatigue analyses, the corresponding equation would be analogous [14]:

$$\Delta\sigma\left(\frac{L}{2}\right) = \Delta\sigma_0 \tag{4}$$

The TCD parameters (the critical distance, L, and the inherent strength, σ_0) can be easily derived by performing two fracture (or fatigue) tests on two specimens presenting different notch radii. At fracture (or fatigue), the stress-distance curves cross each other at one point with coordinates ($L/2$, σ_0) (($L/2$, $\Delta\sigma_0$) in fatigue analysis).

When combined with the stress distribution on the notch tip provided by Creager and Paris [24] (Equation (5)), the PM generates predictions of the apparent fracture toughness (K^N_{mat}) exhibited by materials containing U-shaped notches:

$$K^N_{mat} = K_{mat} \frac{\left(1 + \frac{\rho}{L}\right)^{3/2}}{\left(1 + \frac{2\rho}{L}\right)} \tag{5}$$

2.1.2. The Line Method (LM)

The LM assumes that fracture takes place when the average stress along a distance equal to $2L$ (from the notch tip) reaches the inherent strength, σ_0 [14]:

$$\frac{1}{2L} \int_0^{2L} \sigma(r)dr = \sigma_0 \tag{6}$$

Analogously, in fatigue analysis, the LM criterion is defined by the following equation [14]:

$$\frac{1}{2L} \int_0^{2L} \Delta\sigma(r)dr = \Delta\sigma_0 \tag{7}$$

In combination with the Creager-Paris stress distribution, the LM also provides an estimation of the material apparent fracture toughness in the presence of U-shaped notches [14]:

$$K^N_{mat} = K_{mat} \sqrt{1 + \frac{\rho}{4L}} \tag{8}$$

2.2. Environmentally Assisted Cracking and Hydrogen Embrittlement

Environmentally assisted cracking groups together a wide range of cracking phenomena that take place under aggressive environments. Failure occurs as a consequence of a synergetic action of material, environment and stresses [2]. These phenomena may lead to subcritical crack growth processes and final fracture due to the degradation of the mechanical properties of the materials [5,25].

Hydrogen can cause embrittlement in metals by the action of atoms penetrating the material microstructure and diffusing to the most stressed zones. In order for HE failure to occur, a susceptible material, an exposure to a hydrogen-containing environment and high enough stresses are required. HE can be caused by cathodic polarization, which is a technique employed to prevent corrosion processes by reducing the corrosion rate when a potential (or current density) below the open circuit potential is applied (by means of a cathodic polarization) between the anode and the cathode, usually provided by an external source [8]. However, this method increases the H_2 production, and, if the polarization is excessive, a direct reduction of H_2O is possible, as shown below:

$$2H_2O + 2e^- \rightarrow H_2 + 2OH^- \tag{9}$$

Before the H_2 molecule formation, H atoms are present on the metal surface during a significant time, which is increased by the presence of poisons of the hydrogen recombination reaction (e.g., H_2S and As). Hydrogen atoms can penetrate into interstitial sites, facilitated by their small size, and cause embrittlement [26,27].

3. Materials and Methods

3.1. Materials

This study analyzes, through the application of the PM and the LM, the effect of the environment on the EAC behavior of two steels: a weldable thermo-mechanically treated S420 medium-strength steel [28], and API X80 medium-strength steel obtained by means of control rolling and accelerate cooling [29]. The mechanical properties of these steels, as received, are shown in Table 1:

Table 1. Mechanical properties of the materials being analyzed.

Material	E (GPa)	σ_y (MPa)	σ_u (MPa)	e_u (%)
S420	206.4	447.7	547.1	21.7
0	209.9	621.3	692.9	29.6

The S420 steel is mainly used in offshore structures, pressure vessels and power plants. It presents a ferritic-pearlitic microstructure with a grain size ranging between 5–25 µm. The X80 steel, which is employed in oil and gas transportation at low temperatures, presents a ferritic-pearlitic microstructure with a grain size ranging between 5–15 µm. It is interesting to notice the presence of a small volumetric fraction of bainite/degenerated pearlite and the absence of acicular ferrite in X80 steel. The chemical composition and the microstructure of these steels are gathered in Table 2 and Figure 2, respectively.

Table 2. Chemical composition of the two steels being analyzed.

Material	C	Si	S	P	Mn	Ni	Cr	Mo	Cu	Al	V	Ti	Nb
S420	0.08	0.28	0.001	0.012	1.44	0.03	0.02	0.003	0.015	0.036	0.005	0.015	0.031
X80	0.07	0.18	<0.005	<0.005	1.83	0.03	-	0.15	0.02	0.03	-	-	0.03

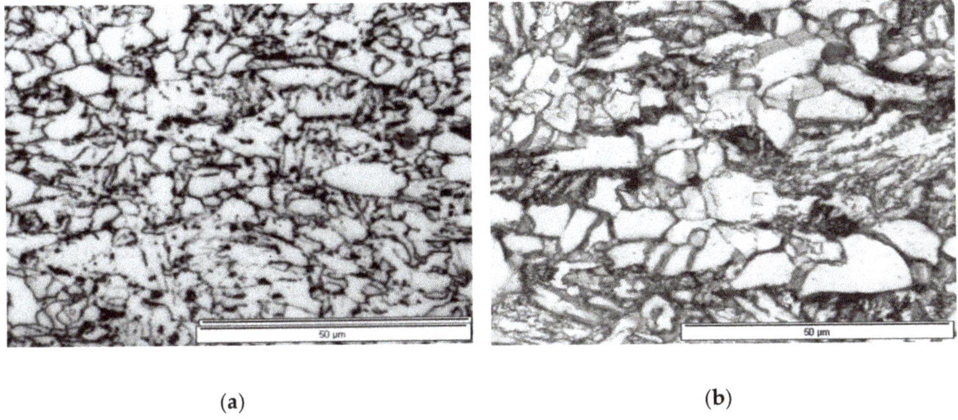

(a)   (b)

Figure 2. Microstructure of: (**a**) X80 steel; (**b**) S420 steel.

3.2. Simulation of Hydrogen Embrittlement

Cathodic polarization (or cathodic charge) has been used in this work to cause embrittlement on the steels through fixed current intensities between the steel and the anode (in this case platinum) [26]. Both metals are connected through an aqueous solution, which is prepared following Pressouyre's method [30] and consists of an 1 N H_2SO_4 solution in distilled water with 10 mg of an As_2O_3 solution and 10 drops of CS_2 per liter of dissolution. The pH of the aqueous environment is kept in the range 0.65–0.80 at room temperature [31,32].

In this study, two levels of current intensity (5 mA/cm^2 and 1 mA/cm^2) have been considered. The former was previously applied in [20], whereas the latter has been specifically applied for this study. Cathodic polarization at these levels is used to cause two levels of hydrogen embrittlement on the steels. These testing conditions have been used, in combination with slow strain rate conditions and notched geometries, to generate hydrogen embrittlement. The application of cathodic polarization in actual structures, and their corresponding structural materials, is generally far from these circumstances, so that HE is avoided.

Figure 3 shows a schematic of the cathodic charge used in this work. The solution is stirred to avoid hydrogen bubbles on the specimen surface and prevent localized corrosion. As shown in Figure 4, the C(T) specimen is used as the working electrode, the platinum grid is used as the counter electrode and the saturated calomel electrode becomes the reference electrode.

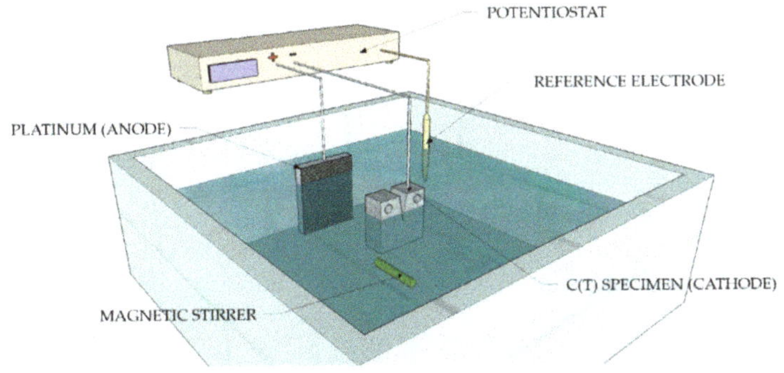

Figure 3. Cathodic polarization employed in this work.

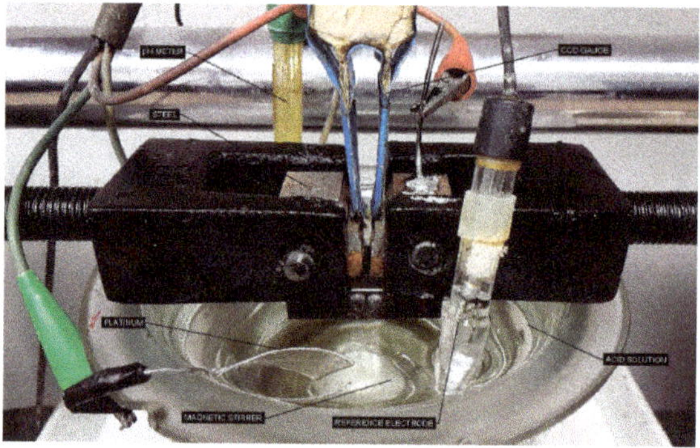

Figure 4. Experimental setup.

3.3. Analysis of Environmental Assisted Cracking: An Approach from the Theory of Critical Distances

The TCD has been reformulated to address EAC processes [20]. In order to obtain the critical distance in EAC conditions, L_{EAC}, the equation proposed is [20]:

$$L_{EAC} = \frac{1}{\pi}\left(\frac{K_{IEAC}}{\sigma_{0EAC}}\right)^2 \tag{10}$$

where L_{EAC} is the EAC critical distance, K_{IEAC} is the EAC crack propagation threshold (obtained from cracked specimens) and σ_{0EAC} is the inherent strength under EAC conditions, which has to be calibrated.

Analogously to fracture processes, and assuming Equation (10) for the EAC critical distance, predictions of the apparent crack propagation threshold for EAC (K^N_{IEAC}), above which EAC initiates and grows from U-shaped notches, can be derived. The combination of Creager-Paris stress distribution [24] with the PM and the LM leads to Equations (11) and (12), respectively:

$$K^N_{IEAC} = K_{IEAC}\frac{\left(1 + \frac{\rho}{L_{EAC}}\right)^{3/2}}{\left(1 + \frac{2\rho}{L_{EAC}}\right)} \tag{11}$$

$$K^N_{IEAC} = K_{IEAC}\sqrt{1 + \frac{\rho}{4L_{EAC}}} \tag{12}$$

These equations allow the apparent EAC crack propagation threshold, K^N_{IEAC}, to be calculated from the notch radius, ρ, the material EAC critical distance, L_{EAC}, and the material EAC crack propagation threshold, K_{IEAC} (calculated in cracked conditions following the methodology proposed in the standard ISO 7539 [33]).

3.4. Experimental Methods

In this work, fatigue pre-cracked C(T) specimens and notched C(T) specimens were tested using a slow strain rate machine. A constant displacement rate of $6 \cdot 10^{-8}$ m/s was employed in each material and environment. Cathodic polarizations at 5 mA/cm^2 and 1 mA/cm^2 were used to generate two different conditions. Both S420 and X80 specimens were manufactured in TL orientation. Standard geometry [33] was used for pre-cracked specimens, slightly modified for notched ones, as shown in Figure 5. As mentioned above, the use of the TCD in EAC analyses has been previously presented in [20]. The resulting methodology was validated in steels S420 and X80 subjected to cathodic polarization at 5 mA/cm^2, obtaining satisfactory results. However, before extending the use of the TCD in EAC assessments, it is necessary to provide further validation. Such validation is provided in this work through the application of the TCD in the analysis of the EAC generated by an additional cathodic polarization (1 mA/cm^2).

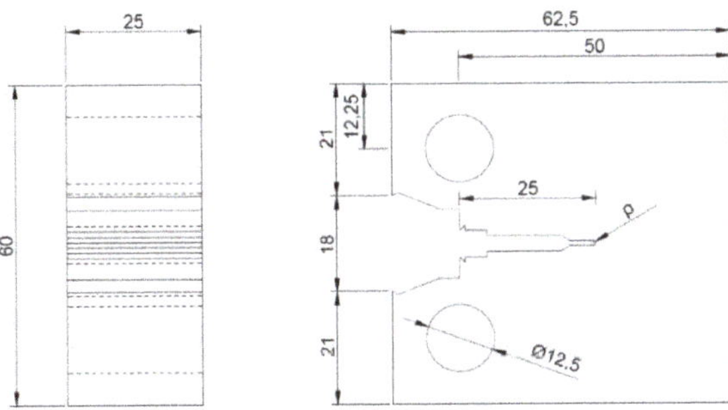

Figure 5. Geometry of the tested specimens (dimensions in mm).

The whole experimental program is composed of 40 C(T) specimens with notch radii varying from 0 mm (crack-like defect) up to 2.0 mm, as shown in Table 3. Half of the tests (those corresponding to cathodic polarization at 5 mA/cm^2) have been previously reported in [20], whereas the other half are totally new. Each test has been duplicated in order to obtain more representative results. Before conducting EAC tests, the specimens were exposed to the aqueous environment under cathodic polarization conditions for 2 days in order to achieve the highest possible level of hydrogen absorption through the corresponding environment. During the hydrogen absorption and the EAC test, the specimens, were submerged in the aqueous solution, ensuring that the defect tip (crack or notch) is always covered by the solution [34]. After the hydrogen absorption, a slow strain rate machine was employed for the mechanical tests, where the specimens were subjected to a constantly rising displacement at $6 \cdot 10^{-8}$ m/s constant displacement rate, while being exposed to the cathodic polarization [35]. Load-COD (Crack Opening Displacement) curves were obtained for all the specimens.

Table 3. Experimental program.

Material	Displacement Rate (m/s)	Cathodic Polarization (mA/cm^2)	ρ (mm)	Number of Tests
			0.00	2
			0.25	2
		5 [20]	0.50	2
			1.00	2
X80	$6 \cdot 10^{-8}$		2.00	2
			0.00	2
			0.25	2
		1	0.50	2
			1.00	2
			2.00	2
			0.00	2
			0.25	2
		5 [20]	0.50	2
			1.00	2
S420	$6 \cdot 10^{-8}$		2.00	2
			0.00	2
			0.25	2
		1	0.50	2
			1.00	2
			2.00	2

The methodology proposed by the standard ISO 7539 [33] was employed in order to calculate the stress intensity factor above which EAC initiates, K_{IEAC}. The corresponding equation is:

$$K_{IEAC} = \frac{P_Q}{(BB_N W)^{1/2}} f\left(\frac{a}{W}\right) \tag{13}$$

where P_Q is the applied load at the propagation onset due to EAC, B is the specimen thickness, B_N is the net specimen thickness ($B = B_N$ if no side grooves are present), W is the specimen width and $f(a/W)$ is a geometrical factor depending on the crack size, a, and the specimen width, W. In case of C(T) specimens, the geometrical factor follows this equation:

$$f\left(\frac{a}{W}\right) = \frac{\left[\left(2 + \frac{a}{W}\right)\left(0.886 + 4.64\frac{a}{W} - 13.32\left(\frac{a}{W}\right)^2 + 14.72\left(\frac{a}{W}\right)^3 - 5.6\left(\frac{a}{W}\right)^4\right)\right]}{\left(1 - \frac{a}{W}\right)^{3/2}} \tag{14}$$

These equations were applied to both cracked specimens (generating K_{IEAC} values) and notched specimens (generating K^N_{IEAC} values).

Finally, a finite element (FE) analysis was carried out in Abaqus (SIMULIA Academic Research Suite, Abaqus 2016) in order to obtain the stress field at the defect tip when the crack starts to propagate. The simulation was performed in linear elastic conditions. The structured meshing technique was used and the model was developed using C3D8R 3D solid elements with reduced integration. This 3D model required manual partitioning, which has been performed following the specimen shape in order to refine the mesh in the most complex regions, such as the notch tip and the holes. The notch tip presents 60 elements around the perimeter and 30 elements along the width. As the distance moves away from the notch tip, the mesh becomes thicker. In other words, the mesh, which has been built using hexahedric elements, is more refined close to the notch tip. Figure 6 represents the FE model employed together with the mesh.

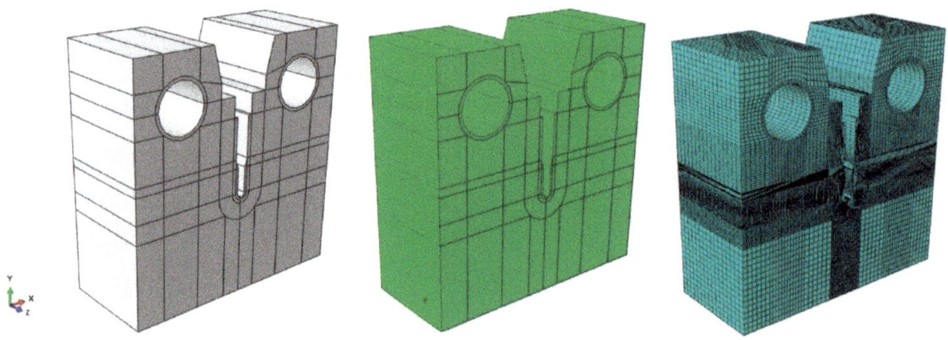

Figure 6. Manual partitioning and mesh employed in the FEA.

4. Results

Table 4 gathers the experimental results of the tests. The values of K^N_{IEAC} have been obtained following the methodology proposed in ISO 7539 and using Equations (13) and (14). For cracked specimens ($\rho = 0$ mm), K^N_{IEAC} coincides with K_{IEAC}.

Figure 7 shows, as an example, four of the experimental Load-COD curves obtained. They all correspond to steel X80, with two different notch radii ($\rho = 0$ mm and $\rho = 2.0$ mm), and the two cathodic polarizations (5 mA/cm^2 and 1 mA/cm^2). A clear notch effect and a much more moderate (but still significant) effect of the cathodic polarization level can be observed.

Table 4. Summary of the experimental results analyzed in this paper.

Material	Specimen	Cathodic Polarization (mA/cm^2)	ρ (mm)	P_Q (kN)	K^N_{IEAC} (MPa·m$^{0.5}$)
X80	X80-5-1	5 [20]	0.00	27.86	67.42
	X80-5-2			23.12	53.16
	X80-5-3		0.25	34.41	63.21
	X80-5-4			34.85	64.01
	X80-5-5		0.50	38.26	70.28
	X80-5-6			42.75	78.52
	X80-5-7		1.00	44.50	81.74
	X80-5-8			42.93	78.87
	X80-5-9		2.00	56.24	103.31
	X80-5-10			54.20	99.59
	X80-1-1	1	0.00	24.30	60.00
	X80-1-2			24.32	54.40
	X80-1-3		0.25	30.70	56.40
	X80-1-4			33.93	62.33
	X80-1-5		0.50	33.53	61.58
	X80-1-6			36.25	66.59
	X80-1-7		1.00	37.90	69.62
	X80-1-8			41.49	76.21
	X80-1-9		2.00	51.24	94.12
	X80-1-10			48.48	89.05
S420	S420-5-1	5 [20]	0.00	28.76	67.58
	S420-5-2			24.01	61.25
	S420-5-3		0.25	34.70	63.74
	S420-5-4			33.94	62.34
	S420-5-5		0.50	37.09	68.13
	S420-5-6			34.21	62.84
	S420-5-7		1.00	41.09	75.48
	S420-5-8			40.49	74.37
	S420-5-9		2.00	45.45	83.49
	S420-5-10			45.67	83.89
	S420-1-1	1	0.00	24.89	62.17
	S420-1-2			24.65	61.61
	S420-1-3		0.25	39.36	72.31
	S420-1-4			36.57	67.17
	S420-1-5		0.50	41.21	75.70
	S420-1-6			39.68	72.90
	S420-1-7		1.00	44.94	82.54
	S420-1-8			43.86	80.58
	S420-1-9		2.00	47.13	86.57
	S420-1-10			46.60	85.60

The stress-distance curves at the notch tip and at crack propagation onset were obtained through the FE model (i.e., when applying the corresponding P_Q). The PM postulates (Figure 1) that the curves cross each other at one point with coordinates ($L_{EAC}/2$, σ_{0EAC}). Figures 8–11 represent the stress-distance curves for the different combinations of steel (X80 and S420) and environmental condition (5 mA/cm^2

and 1 mA/cm^2). The resulting stress-distance curves were obtained with the average value of P_Q obtained in the two tests performed in nominally identical conditions.

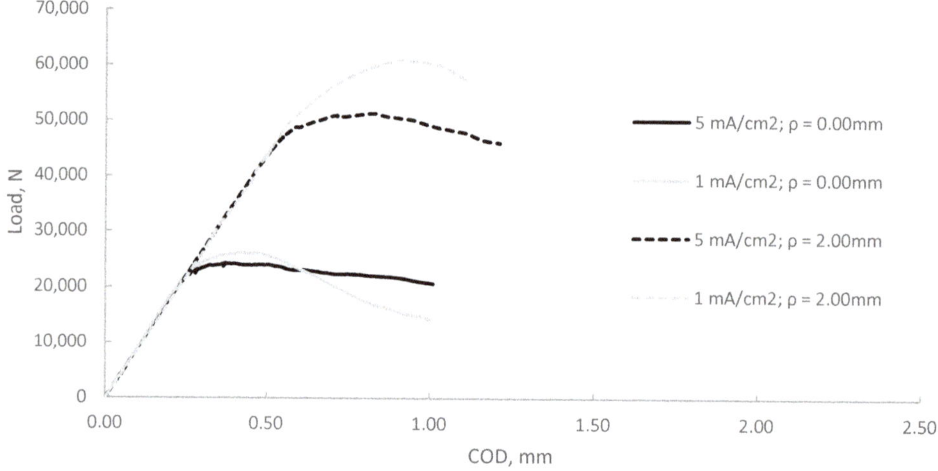

Figure 7. Examples of Load-COD curves (X80 steel, ρ = 0.00 mm and ρ = 2.00 mm).

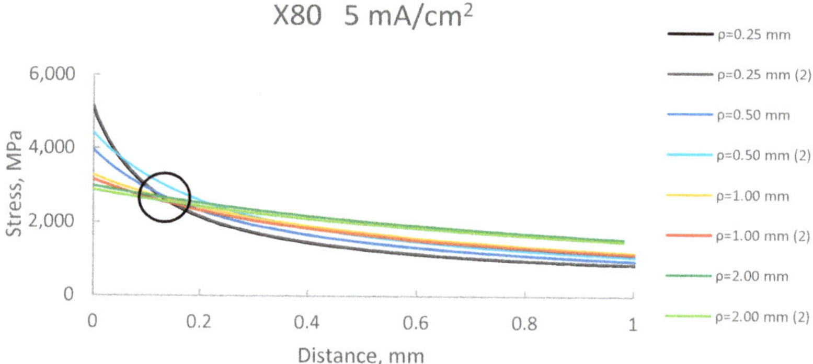

Figure 8. Stress-distance curves at crack initiation in X80 steel at 5 mA/cm^2 of cathodic polarization [20].

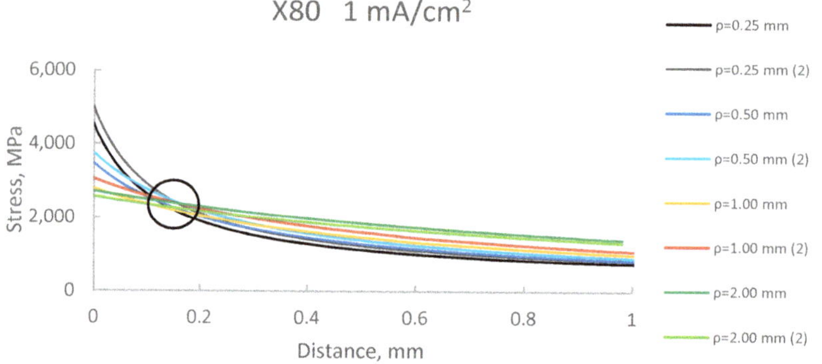

Figure 9. Stress-distance curves at crack initiation in X80 steel at 1 mA/cm^2 of cathodic polarization.

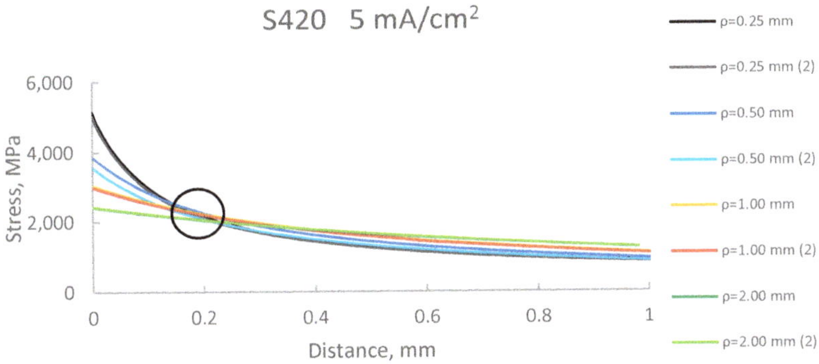

Figure 10. Stress-distance curves at crack initiation in S420 steel at 5 mA/cm^2 of cathodic polarization [20].

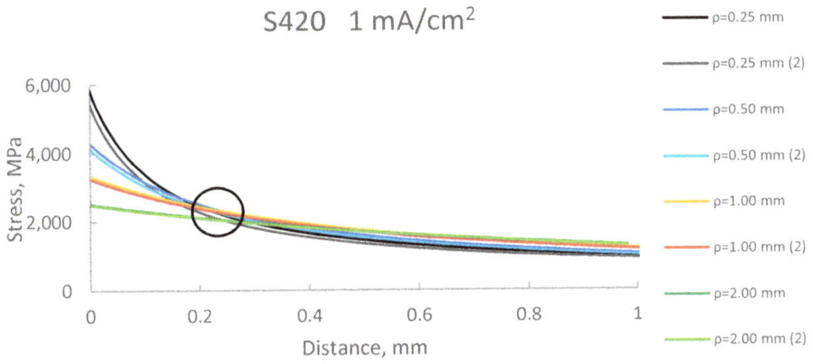

Figure 11. Stress-distance curves at crack initiation in S420 steel at 1 mA/cm^2 of cathodic polarization.

Table 5 shows the experimental results of K_{IEAC} (obtained from Equation (13) and the P_Q values obtained in cracked specimens) and the experimental values of L_{EAC}. Figures 8–11 reveal how the different curves do not cross each other at the same point, although the fundamental assumption of the PM is fulfilled. Taking the average values, obtained in the different cut-off points obtained from the stress-distance curves, the derivation of L_{EAC} is straightforward.

Table 5. Values of K_{EAC}, L_{EAC} and the best fit of the EAC critical distance depending on the methodology.

Material	Cathodic Polarization (mA/cm^2)	K_{IEAC} (MPa·m$^{0.5}$)	L_{EAC} (mm)	PM $L_{EAC\text{-}BF}$ (mm)	LM $L_{EAC\text{-}BF}$ (mm)
X80	5 [20]	60.29	0.286	0.194	0.276
	1	58.37	0.318	0.208	0.303
S420	5 [20]	64.42	0.462	0.386	0.776
	1	61.89	0.499	0.273	0.441

Once K_{IEAC} and L_{EAC} were obtained, predictions of K^N_{IEAC} can be derived for each notch radius following the PM (Equation (11)) or the LM (Equation (12)). Predictions provided by PM and LM using the obtained L_{EAC} values, and their comparison with the corresponding best fit of the experimental results are shown in Figures 12–15. The values of the critical distance providing the best fit (least squares) of the experimental results ($L_{EAC\text{-}BF}$), when using both the PM (Equation (11)) and the LM (Equation (12)), are also gathered in Table 5.

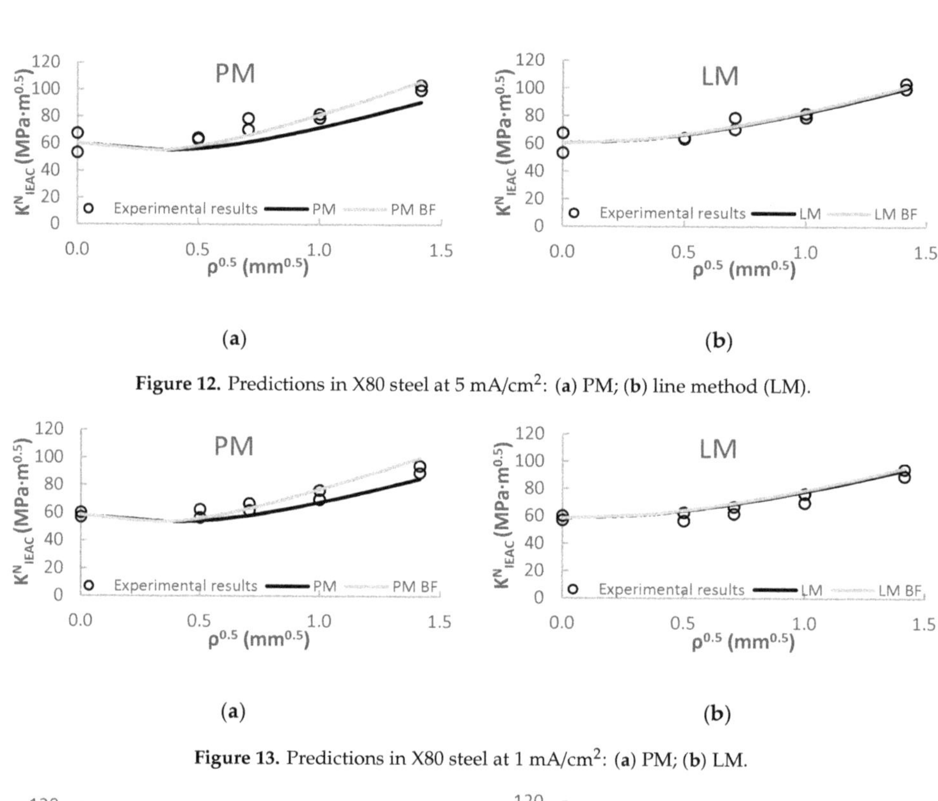

(a)   (b)

Figure 12. Predictions in X80 steel at 5 mA/cm^2: (**a**) PM; (**b**) line method (LM).

(a)   (b)

Figure 13. Predictions in X80 steel at 1 mA/cm^2: (**a**) PM; (**b**) LM.

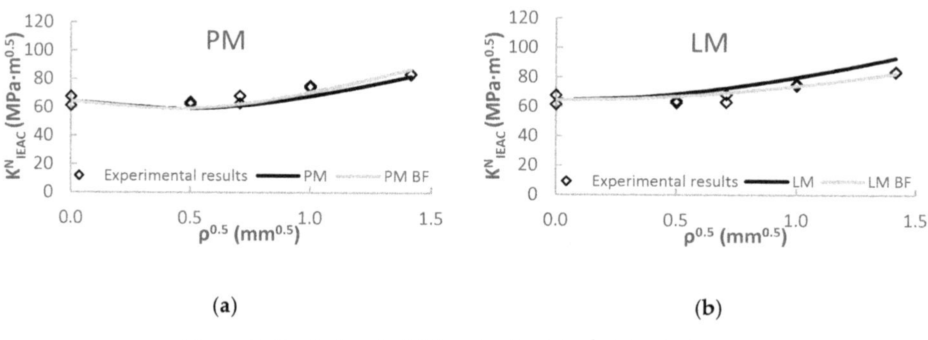

(a)   (b)

Figure 14. Predictions in S420 steel at 5 mA/cm^2: (**a**) PM; (**b**) LM.

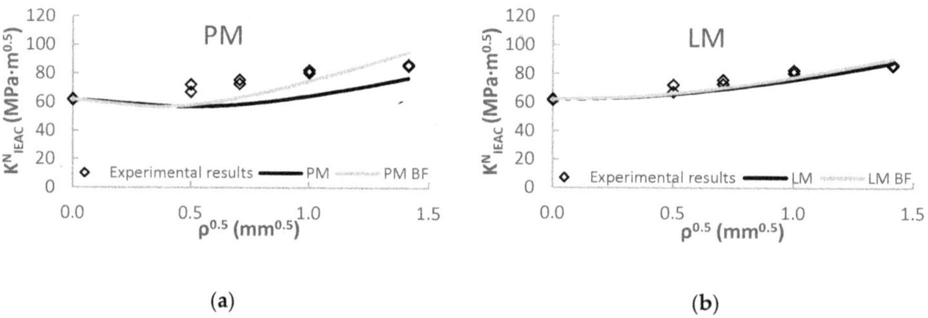

(a)   (b)

Figure 15. Predictions in S420 steel at 1 mA/cm^2: (**a**) PM; (**b**) LM.

5. Discussion

K_{IEAC} values have been obtained as the average value obtained in the pre-cracked specimens ($\rho = 0$ mm), for each combination of material and environment. Even though the S420 steel does not fulfil the recommendation for size independency specified by the standard, the aggressive environment (hydrogen embrittlement) reduces the plasticity and, hence, the sample dimensions required to limit plastic deformations. Consequently, a minimum thickness cannot be specified [33]. In any case, regardless of whether or not K_{IEAC} is size independent for the specimens being tested, the analysis performed allows the results obtained for the different notch radii to be compared (as the thickness is the same for the different specimens). Therefore, and for the sake of simplicity, they are all named equally, without any reference to any possible size independence.

Moreover, the TCD assumes linear-elastic conditions, although the real situation may have certain non-linearities. This may be done, and has been widely validated, once σ_0 (or σ_{0EAC}) is conveniently calibrated. In other words, under non-linear conditions, σ_0 becomes a parameter that converts a non-linear situation into an equivalent linear-elastic one [14,36].

Tests under two levels of cathodic polarization (embrittlement levels) have been carried out in X80 and S420 steels. Stress-distance curves have been obtained through finite element simulations using the load when the crack starts to propagate. In the four combinations of material and environment, the different curves cross each other at approximately one point with coordinates ($L_{EAC}/2, \sigma_{0EAC}$), as the TCD (PM) assumption postulates. Thus, it can be stated that the TCD assumptions are fulfilled in EAC conditions, likewise in fracture and fatigue phenomena.

The values of L_{EAC} at 1 mA/cm^2 of cathodic polarization are larger than those values obtained at 5 mA/cm^2. This is related with the presence of sulfur in the acidic solution, given that sulfur ions acting as poison are less active than the undissociated H_2S during the hydrogen recombination reaction. The test conditions may cause alkalinisation, and this phenomenon increases with the current density. The alkalinisation reduces the hydrogen evolution and may form an oxide layer on the specimen that prevents the hydrogen entry into the steel.

The experimental results of K^N_{IEAC} show that an evident notch effect has been observed in all combinations of material and environment, this effect being more pronounced in the X80 steel. This also implies that steel X80 has lower values of L_{EAC} than steel S420 (it is thus more sensitive to notches).

In any case, it has been observed that both PM and LM provide accurate predictions of K^N_{IEAC}, regardless of the material, environment and notch radius being analyzed. Some of the predictions basically provide the same values as those generated by the best-fit. However, the LM provides predictions that are more accurate, and the PM offers both straightforward and safe predictions.

6. Conclusions

In this paper, the effect of an aggressive environment on the environmentally assisted cracking behavior of two steels containing U-shaped notches has been analyzed using the theory of critical distances (TCD). The TCD has been applied through the point method and the line method to predict the apparent crack propagation threshold for EAC, K^N_{IEAC}, under two hydrogen embrittlement conditions caused by cathodic polarization at a level of 5 mA/cm^2 (previously reported in the literature) and 1 mA/cm^2. TCD parameters have been obtained by a combination of experimental tests and finite element simulations.

Both materials present an evident notch effect, with an increase of K^N_{IEAC} with the notch radius, ρ. This has been accurately predicted using the TCD (PM and LM).

Also, the capacity of the TCD for predicting the effect of the embrittlement conditions in the EAC behavior of the notched steels being analyzed has been demonstrated.

Thus, this work provides additional validation of the use of the TCD for the analysis of EAC processes.

Author Contributions: S.C. and J.A.A. proposed the model, S.C., B.A. and J.A.A. conceived and designed the experiments; P.G. performed the experiments and wrote the paper; all the authors analyzed the data and provided comments and corrections to the first draft of the document.

Funding: This research received no external funding.

Acknowledgments: The authors of this work would like to express their gratitude to the Spanish Ministry of Science and Innovation for the financial support of the projects MAT2014-58738-C3-3-R and MAT2014-58443-P developed by the University of Cantabria.

Conflicts of Interest: The authors declare no conflict of interest. The funders had no role in the design of the study; in the collection, analyses, or interpretation of data; in the writing of the manuscript, or in the decision to publish the results.

References

1. *International Energy Outlook 2016*; U.S. Energy Information Administration: Washington, DC, USA, 2016; ISBN 9781613241431.
2. Wolski, K. Environmentally assisted cracking (EAC) in nuclear reactor systems and components. In *Nuclear Corrosion Science and Engineering*; Woodhead Publishing Series in Energy; Woodhead Publishing Limited: Sawston, UK, 2012.
3. Lynch, S.P. Failures of structures and components by metal-induced embrittlement. In *Stress Corrosion Cracking*; Raja, V.S., Shoji, T., Eds.; Woodhead Publishing Series in Metals and Surface Engineering; Woodhead Publishing: Sawston, UK, 2011; pp. 714–748, ISBN 978-1-84569-673-3.
4. Gangloff, R.P. *Hydrogen Assisted Cracking of High Strenght Alloys*; Aluminum co of America Alcoa Center PA Alcoa Technical Center: Charlotesville, VA, USA, 2003.
5. Brown, B.F. *Stress Corrosion Cracking Control Measures*; National Bureau of Standards Monograph 156; American University Washington DC Department of Chemistry: Washington, DC, USA, 1977.
6. Iannuzzi, M. Environmentally assisted cracking (EAC) in oil and gas production. In *Stress Corrosion Cracking*; Raja, V.S., Shoji, T., Eds.; Woodhead Publishing Series in Metals and Surface Engineering; Woodhead Publishing: Sawston, UK, 2011; pp. 570–607, ISBN 978-1-84569-673-3.
7. Koch, G. Cost of corrosion. In *Trends in Oil and Gas Corrosion Research and Technologies*; Woodhead Publishing: Sawston, UK, 2017; pp. 3–30. [CrossRef]
8. Jones, D.A. *Principles and Prevention of Corrosion*; Prentice-Hall, Inc. Simon & Schuster: Upper Saddle River, NJ, USA, 1996; ISBN 0-13-359 993-0.
9. Elboujdaini, M. Hydrogen-Induced Cracking and Sulfide Stress Cracking. In *Uhlig's Corrosion Handbook*, 3rd ed.; John Wiley & Sons, Inc.: Hoboken, NJ, USA, 2011; pp. 183–194, ISBN 9780470080320.
10. Atzori, B.; Lazzarin, P.; Filippi, S. Cracks and notches: Analogies and differences of the relevant stress distributions and practical consequences in fatigue limit predictions. *Int. J. Fatigue* **2001**, *23*, 355–362. [CrossRef]
11. Fenghui, W. Prediction of intrinsic fracture toughness for brittle materials from the apparent toughness of notched-crack specimen. *J. Mater. Sci.* **2000**, *35*, 2543–2546. [CrossRef]
12. Pluvinage, G. Fatigue and fracture emanating from notch; the use of the notch stress intensity factor. *Nucl. Eng. Des.* **1998**, *185*, 173–184. [CrossRef]
13. Nui, L.S.; Chehimi, C.; Pluvinage, G. Stress field near a large blunted tip V-notch and application of the concept of the critical notch stress intensity factor (NSIF) to the fracture toughness of very brittle materials. *Eng. Fract. Mech.* **1994**, *49*, 325–335. [CrossRef]
14. Taylor, D. *The Theory of Critical Distances: A New Perspective in Fracture Mechanics*; Elsevier: Oxford, UK, 2007; ISBN 9780080444789.
15. Taylor, D.; Kasiri, S.; Brazel, E. The theory of critical distances applied to problems in fracture and fatigue of bone. *Frat. Integrità Strutt.* **2009**, *3*. [CrossRef]
16. Susmel, L. The theory of critical distances: A review of its applications in fatigue. *Eng. Fract. Mech.* **2008**, *75*, 1706–1724. [CrossRef]
17. Cicero, S.; Madrazo, V.; Carrascal, I.; García, T. The use of the theory of critical distances in fracture and structural integrity assessments. In *Research and Applications in Structural Engineering, Mechanics and Computation, Proceedings of the 5th International Conference on Structural Engineering, Mechanics and Computation (SEMC 2013), Cape Town, South Africa, 2–4 September 2013*; CRC Press: Boca Raton, FL, USA; pp. 573–578.

18. Ibáñez-Gutiérrez, F.T.; Cicero, S.; Carrascal, I.A. Analysis of notch effect in short glass fibre reinforced polyamide 6. In Proceedings of the 17th European Conference on Composite Materials (ECCM 2016), Munich, Germany, 26–30 June 2016.

19. Cicero, S.; Madrazo, V.; Carrascal, I.A.; Cicero, R. Assessment of notched structural components using Failure Assessment Diagrams and the Theory of Critical Distances. *Eng. Fract. Mech.* **2011**, *78*, 2809–2825. [CrossRef]

20. González, P.; Cicero, S.; Arroyo, B.; Álvarez, J.A. A Theory of Critical Distances based methodology for the analysis of environmentally assisted cracking in steels. *Eng. Fract. Mech.* **2019**. [CrossRef]

21. Neuber, H. *Theory of Notch Stresses: Principles for Exact Calculation of Strength with Reference to Structural form and Material*; Springer Verlag: Berlin, Germany, 1958; Volume 4547.

22. Peterson, R.E. Notch sensitivity. In *Metal Fatigue*; McGraw Hill: New York, NY, USA, 1959; pp. 293–306.

23. Cicero, S.; Fuentes, J.D.; Procopio, I.; Madrazo, V.; González, P. Critical distance default values for structural steels and a simple formulation to estimate the apparent fracture toughness in u-notched conditions. *Metals* **2018**, *8*. [CrossRef]

24. Creager, M.; Paris, P.C. Elastic field equations for blunt cracks with reference to stress corrosion cracking. *Int. J. Fract. Mech.* **1967**, *3*, 247–252. [CrossRef]

25. Gangloff, R.P. Hydrogen-assisted Cracking. *Compr. Struct. Integr.* **2003**, *6*, 31–101. [CrossRef]

26. Hamilton, J.M. The challenges of deep-water arctic development. *Int. J. Offshore Polar Eng.* **2011**, *21*, 241–247.

27. Shipilov, S.; Jones, R.; Olive, J.-M.; Rebak, R. *Environment-Induced Cracking of Materials*; Shipilov, S.A., Jones, R.H., Olive, J.-M., Rebak, R.B., Eds.; Elsevier: Amsterdam, The Netherlands, 2008; ISBN 9780080446356.

28. *BS EN 10225:2009, Weldable Structural Steels for Fixed Offshore Structures Technical Delivery Conditions*; BSI: London, UK, 2009; ISBN 9780580646232.

29. Specification, A.P.I. *5LD-2009: Specification for CRA Clad or Lined Steel Pipe*; American Petroleum Institute: Washington, DC, USA, 2009.

30. Bernstein, I.M.; Pressouyre, G.M. *Role of Traps in the Microstructural Control of Hydrogen Embrittlement of Steels*; Noyes Publ: Park Ridge, NJ, USA; Pittsburgh, PA, USA, 1988; ISBN 0815510276.

31. Arroyo, B.; Álvarez, J.A.; Lacalle, R.; Uribe, C.; García, T.E.; Rodríguez, C. Analysis of key factors of hydrogen environmental assisted cracking evaluation by small punch test on medium and high strength steels. *Mater. Sci. Eng. A* **2017**, *691*, 180–194. [CrossRef]

32. Alvarez, J.A.; Gutiérrez-Solana, F. Elastic-plastic fracture mechanics based methodology to characterize cracking behavior and its application to environmental assisted processes. *Nucl. Eng. Des.* **1999**, *188*, 185–202. [CrossRef]

33. *ISO 7539 Corrosion of Metals and Alloys. Stress Corrosion Testing. Parts 1 to 11*; ISO: Geneva, Switzerland, 2015.

34. Dietzel, W.; Srinivasan, P.B.; Atrens, A. Testing and evaluation methods for stress corrosion cracking (SCC) in metals. In *Stress Corrosion Cracking*; Raja, V.S., Shoji, T., Eds.; Woodhead Publishing Series in Metals and Surface Engineering; Woodhead Publishing: Sawston, UK, 2011; pp. 133–166, ISBN 978-1-84569-673-3.

35. Turnbull, A. Test methods for environment assisted cracking. *Br. Corros. J.* **1992**, *27*, 271–289. [CrossRef]

36. Susmel, L.; Taylor, D. An Elasto-Plastic Reformulation of the Theory of Critical Distances to Estimate Lifetime of Notched Components Failing in the Low/Medium-Cycle Fatigue Regime. *J. Eng. Mater. Technol.* **2010**, *132*, 021002. [CrossRef]

Article

Finite Fracture Mechanics Assessment in Moderate and Large Scale Yielding Regimes

Ali Reza Torabi [1], Filippo Berto [2] and Alberto Sapora [3,*]

[1] Fracture Research Laboratory, Faculty of New Science and Technologies, University of Tehran, P.O. Box 14395-1561 Tehran, Iran; a_torabi@ut.ac.ir

[2] Department of Industrial and Mechanical Engineering, Richard Birkelands vei 2b, Norwegian University of Science and Technology, 7034 Trondheim, Norway; filippo.berto@ntnu.no

[3] Department of Structural, Building and Geotechnical Engineering, Politecnico di Torino, Corso Duca degli Abruzzi 24, 10129 Torino, Italy

* Correspondence: alberto.sapora@polito.it; Tel.: +39-011-090-4819

Received: 23 April 2019; Accepted: 22 May 2019; Published: 24 May 2019

Abstract: The coupled Finite Fracture Mechanics (FFM) criteria are applied to investigate the ductile failure initiation at blunt U-notched and V-notched plates under mode I loading conditions. The FFM approaches are based on the simultaneous fulfillment of the energy balance and a stress requirement, and they involve two material properties, namely the fracture toughness and the tensile strength. Whereas the former property is obtained directly from experiments, the latter is estimated through the Equivalent Material Concept (EMC). FFM results are presented in terms of the apparent generalized fracture toughness and compared with experimental data already published in the literature related to two different aluminium alloys, Al 7075-T6 and Al 6061-T6, respectively. It is shown that FFM predictions can be accurate even under moderate or large scale yielding regimes.

Keywords: blunt V-notches; aluminium plates; mode I loading; ductile failure; FFM; EMC

1. Introduction

It is well-assessed that brittle fracture initiation in notched structures can be described accurately by combining a linear elastic fracture analysis with an internal material length l. This framework is generally known as Finite Fracture Mechanics (FFM), underlying the fact that the provisional fracture is supposed to proceed through a finite crack extension, at least at crack initiation. Different criteria have been proposed [1] by imposing that either the average energy [2,3] or the punctual/average stress [1,3,4] at a finite distance l from the notch root are greater than the material properties, i.e., the fracture energy G_c ($G_c \propto K_{Ic}^2$, K_{Ic} being the fracture toughness) or the tensile strength σ_u, respectively. The crack advance l results to be a function of K_{Ic} and σ_u, and more in general it becomes a structural parameter in the case of coupled criteria, i.e., when the energy and stress conditions are coupled together [5,6]. It is worthwhile to mention that under the small scale yielding regime, the generalized stress intensity factor (GSIF, governing the asymptotic expansions of both the stress field and the SIF related to a virtual crack stemming from the notch tip providing the crack driving force) results to be the dominating failure parameter: Fracture can thus be supposed to take place when it reaches its critical value, also known as generalized (or notch) fracture toughness.

Undoubtedly, all the FFM approaches are highly sensitive to the material properties, which must be then estimated carefully. Whilst the fracture toughness can be obtained experimentally following the standard ASTM codes, the discussion is open when dealing with the tensile strength. As a matter of fact, the behaviour of plain specimens is strongly affected by the presence of micro-cracks/defects, and the experimental value of σ_u obtained by tensile testing hourglass-shape samples can be lower than the real one σ_0. The ratio σ_0/σ_u can vary sensibly from material to material. Taylor [1] observed by fitting the value of σ_0 through the simple point stress criterion that the ratio σ_0/σ_u approaches the unit value for ceramics, it is comprised between one and two for polymers, and can be sensibly larger (even up to eight) for metals. Of course, in this latter case, it is difficult to refer to σ_0 as a tensile strength, and some attempts to provide a clear physical/mechanics were recently put forward [7]. It should be also mentioned that in order to avoid the problems raised when dealing with plain specimens, Seweryn [3,4] decided to test blunt notched samples where the root radius was large enough to provide a nearly constant stress field: The maximum tensile stress at the notch tip was estimated equal to σ_0.

Since the work by Susmel and Taylor [8], the idea that the above criteria could be applied to deal also with ductile fracture has spread out through the Scientific Community. Susmel and Taylor [8], observing static failures in notched specimens made of a commercial cold-rolled low-carbon steel, found stress criterion predictions to be highly accurate, generally falling within an error interval of about 15%. The parameter σ_0 was estimated through a fitting procedure on the stress criterion, namely through the intersection point of the failure stress fields for a blunt and a sharp notched structure. Finally, it was concluded that prediction accuracy slightly increased by using a numerical elasto-plastic analysis, but not so much to justify the huge computational effort.

Following the same approach and using the same stress criteria, analogous considerations were put forward after tensile testing U-notched samples made of Al 7075-T651 [9]. Indeed a slightly different approach was proposed in [10], where it was suggested, instead of working on the tensile strength, to calibrate the critical length when dealing with structural steels.

More recently, an extensive experimental campaign has been carried out to investigate the ductile fracture of U-notched and blunt V-notched plates [11–14] under mode I loading conditions. The above tests involved two aluminium alloys, Al 7075-T6 and Al 6061-T6, respectively: It was observed that Al 7075-T6 plates failed by moderate scale yielding, whereas Al 6061-T6 plates failed by large scale yielding.

Additionally, in this case, stress criteria were exploited to match experimental data, but σ_0 was evaluated through the Equivalent Material Concept (EMC) theory proposed by Torabi [15]. Once the fracture toughness K_{Ic} is fixed, the EMC idea is to compare the behaviour of a ductile material with that of a virtual brittle material possessing a different tensile strength σ_0. Its values are determined by considering identical values for the strain energy density (i.e., the area below the stress-strain curve) required by the ductile material under investigation and by the virtual brittle one. In formulae:

$$\sigma_0 = \left[\sigma_Y^2 + \frac{2EK}{n+1}\left(\varepsilon_u^{n+1} - 0.002^{n+1}\right)\right]^{0.5} \tag{1}$$

where σ_Y is the yield strength, E is the Young's modulus, K is the strain-hardening coefficient, n is the strain-hardening exponent, and ε_u is the true plastic strain at maximum load. Thus, differently from the previous approaches where σ_0 was no more than a fitting parameter, according to EMC it recovers a precise physical meaning, and its value can be determined starting from the real experimental properties. More recently also the strain energy density (SED) criterion [2] has been exploited and applied to the experimental results as before [16,17].

The purpose of the present work is to investigate the mode I behaviour of blunt V-notched plates made of Al 7075-T6 and Al 6061-T6 [11–14] by combining the coupled FFM approaches and EMC. It is important to point out that, with respect to the stress criteria, coupled FFM approaches allow to remove some inconsistencies [6], their extension to complex geometries, such as interfacial cracks [18], results are more straightforward, and they provide close predictions to the Cohesive Zone Model once the constitutive law is defined *ad-hoc* [19–21]. Some recent applications involve the study of stable/unstable crack propagation from a circular hole [22,23], the extension to 3D crack propagation [24], and the study of crack patterns at bi-material interfaces [25,26].

2. FFM Criteria

The coupled FFM approaches are based on the hypothesis of a finite crack extension l and assume the simultaneous fulfillment of two conditions [5,6].

The first FFM condition establishes that the average energy available for a crack length increment l (evaluated by integrating the crack driving force G over l) is higher than the fracture energy G_c. By referring to the blunt V-notch geometry reported in Figure 1 (ρ being the notch root radius), we have:

$$\frac{1}{l}\int_0^l G(c)\,\mathrm{d}c \geq G_c \tag{2}$$

Equation (2), under mode I loading conditions, can be rewritten by means of Irwin's relationship as:

$$\frac{1}{l}\int_0^l K_I^2(c)\,\mathrm{d}c \geq K_{Ic}^2 \tag{3}$$

where $K_I(c)$ is the SIF related to a crack of length c stemming from the notch root (Figure 1). An analytical relationship for the function $K_I(c)$ was proposed by Sapora et al. [27]

$$K_I(c) = K_I^V \beta\, c^{\lambda-1/2}\left\{1+\left[\left(\frac{\beta}{\psi}\right)^{\frac{1}{1-\lambda}}\frac{r_0}{c}\right]^m\right\}^{\frac{\lambda-1}{m}} \tag{4}$$

where K_I^V is the GSIF referring to a V-notch with the same depth, and λ, β, m are functions only of the notch amplitude ω. Their values are reported in Table 1.

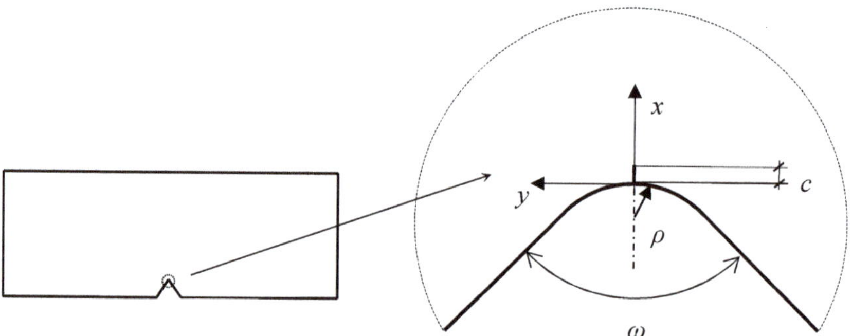

Figure 1. Blunt V-notch geometry. The presence of a "virtual" crack of length c stemming from the notch tip is also depicted.

Table 1. Values of different parameters compared to notch opening angle ω.

ω	λ	β	η	μ	m
0°	0.5000	1.000	1.000	−0.5000	1.820
30°	0.5015	1.005	1.034	−0.4561	1.473
60°	0.5122	1.017	0.9699	−0.4057	1.338
90°	0.5445	1.059	0.8101	−0.3449	1.314

Furthermore, we have that:

$$\psi = 1.12\,\sqrt{\pi}(1+\eta)(2\pi)^{\lambda-1} \tag{5}$$

and

$$r_0 = \frac{2\pi - \omega}{\pi - \omega}\rho \tag{6}$$

Note that the original relationship proposed by Carpinteri et al. [28] did not include the parameter m, which was introduced later [27] to improve the prediction accuracy for low notch amplitudes. The presence of m in (4) does not prevent to fulfill the asymptotic limits for very short and very long (but still small in respect to the notch depth) cracks: Expected errors were estimated below 1% over the range $0 \leq c/\rho \leq 10$ ($0° \leq \omega \leq 180°$).

Introducing Equation (4) into Equation (3) yields:

$$\frac{K_I^V}{K_{Ic}r_0^{1/2-\lambda}} \geq h\left(\bar{l}\right) = \frac{\sqrt{\bar{l}}}{\beta}\left(\int_0^{\bar{l}} \bar{c}^{2\lambda-1}\left\{1 + \left[\left(\frac{\beta}{\psi}\right)^{\frac{1}{1-\lambda}}\frac{1}{\bar{c}}\right]^m\right\}^{\frac{2(\lambda-1)}{m}} d\bar{c}\right)^{-0.5} \tag{7}$$

where $\bar{l} = l/r_0$, and $\bar{c} = c/r_0$.

The second FFM condition involves the circumferential stress $\sigma_y(x)$ together with the material tensile strength σ_u. It can be either a punctual requirement, $\sigma_y(x) \geq \sigma_u$ for $0 \leq x \leq l$, which can be expressed as [5]

$$\sigma_y(x = l) \geq \sigma_u \tag{8}$$

since the stress field is a monotonically decreasing function, or an average condition [6]

$$\frac{1}{l}\int_0^l \sigma_y(x)\,dx \geq \sigma_u \tag{9}$$

By assuming that the notch tip radius ρ is sufficiently small in respect to the notch depth, the stress field along the notch bisector could be expressed as [29]

$$\sigma_y(x) = \frac{K_I^V}{(2\pi(x + r_0))^{1-\lambda}}\left[1 + \eta\left(\frac{r_0}{x + r_0}\right)^{\lambda-\mu}\right] \tag{10}$$

The stress requirement can be easily rewritten by means of Equation (10) as

$$\frac{K_I^V}{\sigma_u r_0^{1-\lambda}} \geq f(\bar{l}) \tag{11}$$

where the function f assumes different forms if the punctual condition (8) is considered

$$f(\bar{l}) = \frac{\left[2\pi\left(1 + \bar{l}\right)\right]^{1-\lambda}}{1 + \eta\left(1 + \bar{l}\right)^{\mu-\lambda}} \tag{12}$$

or if the average condition (9) is taken into account

$$f(\bar{l}) = (2\pi)^{1-\lambda}\bar{l}\left\{\left[\frac{\left(\bar{l}+1\right)^\lambda - 1}{\lambda}\right] + \eta\left[\frac{\left(\bar{l}+1\right)^\mu - 1}{\mu}\right]\right\}^{-1} \tag{13}$$

Let us now suppose that failure takes place when the GSIF K_I^V, reaches its critical conditions $K_{Ic}^{V,\rho}$. Note that $K_{Ic}^{V,\rho}$ represents the apparent generalized fracture toughness (i.e., the generalized fracture toughness measured as if the V-notch was sharp): It depends on the root radius, differently from the GSIF, K_I^V [25–27]. In critical conditions and for positive geometries [22], inequalities (7) and (11) become a system of two equations in two unknowns: The critical crack advancement l_c, and the apparent generalized fracture toughness $K_{Ic}^{V,\rho}$ (i.e., the failure load).

From a more general point of view, by equaling expressions (7) and (11) in critical conditions with respect to $K_{Ic}^{V,\rho}$, the problem can be recast in the following form:

$$\begin{cases} \left[\dfrac{f(\bar{l}_c)}{h(\bar{l}_c)}\right]^2 = \dfrac{1}{r_0}\left(\dfrac{K_{Ic}}{\sigma_u}\right)^2 \\ \dfrac{K_{Ic}^{V,\rho}}{\sigma_u r_0^{1-\lambda}} = f(\bar{l}_c) \end{cases} \tag{14}$$

Thus, for a specified material (K_{Ic}, σ_u) and a given geometry (ω, ρ), the critical crack advancement l_c can be evaluated from the former equation in (14). This value must then be inserted into the latter equation to obtain the apparent generalized fracture toughness $K_{Ic}^{V,\rho}$.

In the following section, Equation (14) will be implemented by distinguishing between "punctual FFM" (if h and f are defined via Equations (7) and (12), respectively) and "average FFM" (if h and f are defined via Equations (7) and (13), respectively). Note that the presence of the parameter m in Equation (7) prevents to get a closed form solution for the integrand function as in [28]: In the present analysis, the integral is computed numerically by applying an adaptive Simpson quadrature formula in a recursive way. The total computational cost related to solve Equation (14) through a simple MATLAB $^{\circledR}$ code is less than 5 s for each geometry, showing the potentiality of the present semi-analytical approach.

3. Comparison with Experimental Results

Let us now consider the results rising from the experimental campaign carried out in [11–14]. Tests involved: I) Two different aluminium alloys, namely Al 7075-T6 and Al 6061-T6; II) four notch amplitudes for each material: $\omega = 0°, 30°, 60°,$ and $90°$; III) three different notch radii for each amplitude: $\rho = 0.5, 1,$ and 2 mm for U-notches ($\omega = 0°$); $\rho = 1, 2,$ and 4 mm for other notch amplitudes. Details about the sample dimensions, the testing procedures, and the related results can be found in the papers quoted above.

In order to apply the FFM criteria described in the previous section, the values for the material properties are required. Whereas the fracture toughness K_{Ic} was derived experimentally, the tensile strength σ_u was obtained through the EMC by means of Equation (1) (and thus the parameter is re-termed as σ_0, Section 1): Table 2 reports the values implemented in the present analysis taken from [12,13]. The ratio σ_0/σ_u is equal to 3.16 for Al 7075-T6 and to 3.65 for Al 6061-T6, reflecting the fact failure of the samples made of Al 6061-T6 involved a larger amount of plasticity.

Table 2. Material properties implemented in the present analysis.

Material	Al 7075-T6	Al 6061-T6
K_{Ic} (MPa $\sqrt{\text{m}}$)	50	38
σ_0 (MPa)	1845	1066

Results in terms of the apparent generalized fracture toughness are presented in Figures 2–5 as concerns $\omega = 0°, 30, 60,$ and $90°$, respectively. It can be noted that punctual FFM [5] always provides higher predictions than average FFM [6], confirming the trend observed in past works (e.g., [25]). Indeed, for what concerns U-notched structures theoretical predictions by both criteria are always satisfactory (Figure 2), the maximum percentage discrepancy being nearly 20%. Punctual FFM results to be more accurate for Al 7075-T6 U-notched plates, whilst average FFM provides the best results for U-notched Al 6061-T6 plates: In this case, the discrepancy is less than 2.3%.

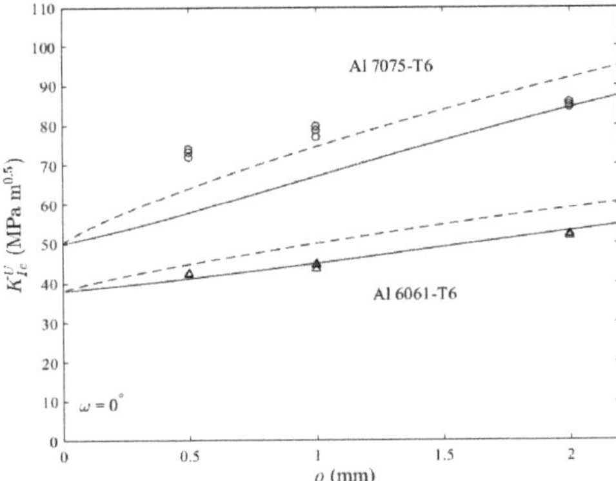

Figure 2. U-notched structures ($\omega = 0°$), apparent fracture toughness: Predictions by punctual FFM (dashed line) and average FFM (continuous line) related to experimental data on Al 7075-T6 plates (circles) and on Al 6061-T6 plates (triangles).

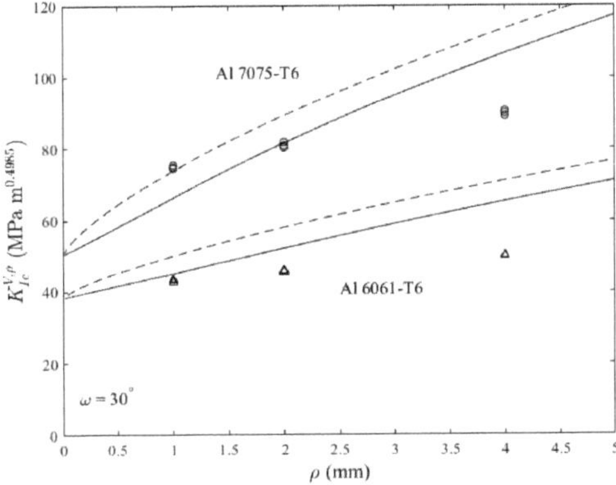

Figure 3. Blunt V-notched structures ($\omega = 30°$), apparent generalized fracture toughness: Predictions by punctual FFM (dashed line) and average FFM (continuous line) related to experimental data on Al 7075-T6 plates (circles) and on Al 6061-T6 plates (triangles).

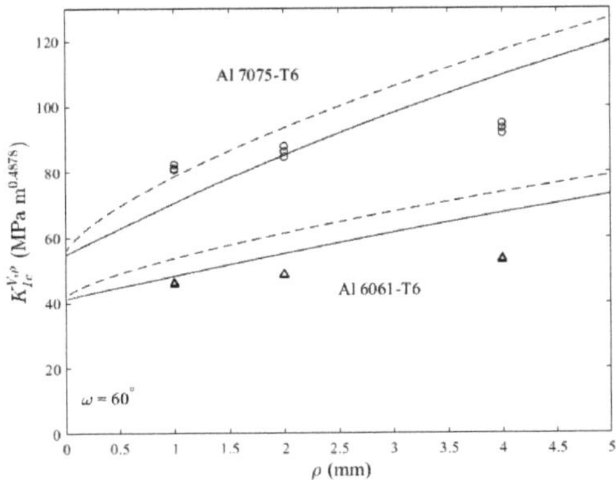

Figure 4. Blunt V-notched structures ($\omega = 60°$), apparent generalized fracture toughness: Predictions by punctual FFM (dashed line) and average FFM (continuous line) related to experimental data on Al 7075-T6 plates (circles) and on Al 6061-T6 plates (triangles).

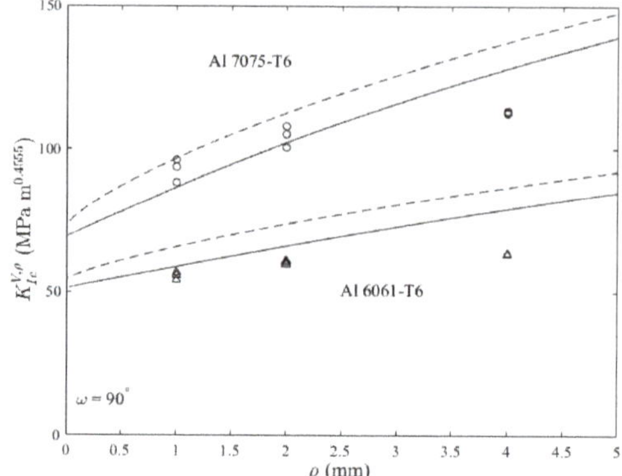

Figure 5. Blunt V-notched structures ($\omega = 90°$), apparent generalized fracture toughness: Predictions by punctual FFM (dashed line) and average FFM (continuous line) related to experimental data on Al 7075-T6 plates (circles) and on Al 6061-T6 plates (triangles).

For blunt V-notches (Figures 3–5), the following considerations hold: (i) The average FFM generally provides the most accurate predictions; (ii) for Al 7075-T6 notched plates, the maximum discrepancy between the average FFM results and experimental data is always below 20%, the maximum discrepancy being observed for large radii; (iii) for Al 6061-T6 notched samples, theoretical predictions are generally less accurate, and they always overestimate the experimental data. The accuracy decreases with the radius, but for this material the discrepancy is much higher for the largest radius ($\rho = 4$ mm), exceeding always 20%, and reaching 30% for $\omega = 30°$. The discrepancy related to the punctual FFM is significantly higher.

The situation for what concerns the percentage discrepancy between experimental results and average FFM predictions is summarized in Figure 6. In order to explain the higher deviation observed for Al 6061-T6 notched alloys with large radii, it should be said that first of all the precision of the asymptotic expansions (4) and (10) decreases as larger radii are considered. Anyway, this does not explain a discrepancy of 30% and the best accuracy of predictions for Al 7075-T6 plates, which possess exactly the same geometry. Thus, it can be supposed that some plastic phenomena took place during the failure mechanism affecting significantly the failure load. In other words, as ρ increases involving higher failure loads, the level of constraint can reduce as plastic zones become larger. This phenomenon is more pronounced for Al 6061-T6 structures, where a large scale yielding was observed, than for Al 7075-T6 plates which failed under moderate scale yielding [11–14]. Ductility is associated to complex failure mechanisms, which the present coupled FFM-EMC model (as well as other approaches recently proposed [8–17]) simply neglects.

Finally, the average FFM diagrams of the apparent generalized fracture toughness and of the critical crack extension compared to the root radius are reported in Figure 7. In order to get dimensionless quantities, the well-known Irwin's length $l_{ch} = (K_{Ic}/\sigma_0)^2$ is introduced. From Table 2, it follows that $l_{ch} \approx 0.73$ mm for Al7075-T6, and $l_{ch} \approx 1.3$ mm for Al6061-T6. It can be seen that the apparent generalized fracture toughness increases as the radius increases and/or the angle decreases. Thus, the cracked configuration is the most affected one by the presence of a radius $\rho \neq 0$. As concerns the crack advance l_c, it results a function of the material (through l_{ch}) and of the radius, tending to $2/(\pi \cdot 1.12^2)$ as ρ tends to infinite (smooth elements): From Figure 7 it can be seen that the values related to the crack case ($\omega = 0°$) are the lowest ones for non-negligible radii. The analytical expressions for K_{Ic}^V and l_c when $\rho = 0$ (sharp case) can be found in [6].

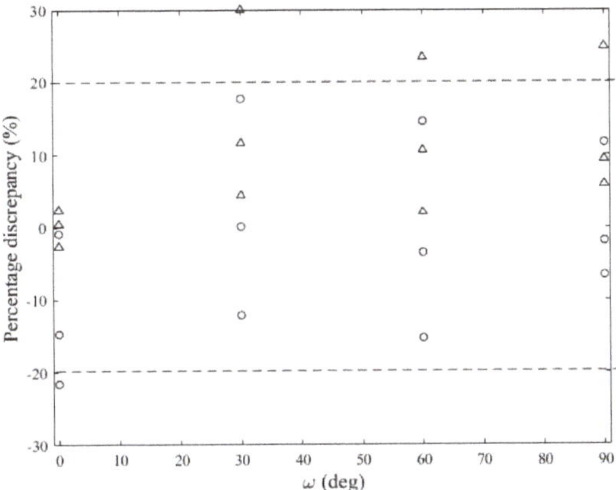

Figure 6. Average FFM predictions: Percentage discrepancy with respect to experimental data referring to blunt V-notched structures made of Al 7075-T6 (circles), and made of Al 6061-T6 (triangles).

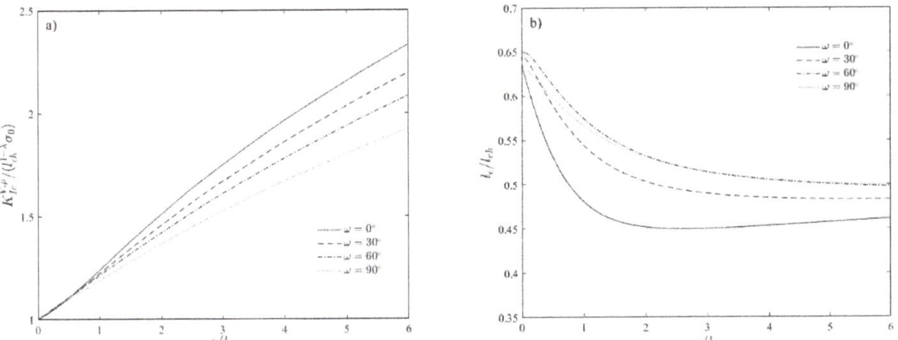

Figure 7. Average FFM: Dimensionless apparent generalized fracture toughness (**a**) and dimensionless critical crack extension (**b**) compared to dimensionless notch root radius.

4. Conclusions

It is consolidated that in order to describe the failure mechanisms of ductile structures, an elasto-plastic finite element analysis has to be carried out. Nevertheless, this procedure generally requires huge computational efforts, and it may not be efficient from an engineer point of view. In this work, in the spirit of the previous papers [12,16], the coupled FFM approaches proposed by Leguillon [5] and Carpinteri et al. [6] are implemented to evaluate the mode I failure load of blunted V-notched samples made of two different materials, Al 7075-T6 and Al 6061-T6. By estimating the tensile strength through the EMC put forward by Torabi [15], it is found that FFM predictions are generally in good agreement with experimental results for Al 7075-T6, especially as regards the average criterion [6]. On the other hand, for Al 6061-T6, FFM results always overestimate the experimental data, becoming mediocre for large radii. In this case, the present coupled FFM-EMC formulation is not able to catch appropriately the effect of the large amount of plasticity involved in the failure mechanisms. Indeed, the EMC approach is formulated based on the standard stress-strain curve of the un-notched specimen, and it does not consider the local plasticity around the notch. Future work will consist of investigating more in details this framework to verify the accuracy of the coupled FFM-EMC model, calibrating its efficiency on the ductility of the structural behavior.

Author Contributions: All the authors conceived the idea of the research and contributed equally to the analysis tools; A.S. wrote the paper.

Funding: This research received no external funding.

Conflicts of Interest: The authors declare no conflict of interest.

References

1. Taylor, D. *The Theory of Critical Distances: A New Perspective in Fracture Mechanics*; Elsevier: Oxford, UK, 2006.
2. Lazzarin, P.; Zambardi, R. A Finite-volume-energy based approach to predict the static and fatigue behaviour of components with sharp V-shaped notches. *Int. J. Fract.* **2002**, *112*, 275–298. [CrossRef]
3. Seweryn, A.; Lukaszewicz, A. Verification of brittle fracture criteria for elements with V-shaped notches. *Engng. Fract. Mech.* **2002**, *69*, 1487–1510. [CrossRef]
4. Seweryn, A. Brittle fracture criterion for structures with sharp notches. *Eng. Fract. Mech.* **1994**, *47*, 673–681. [CrossRef]
5. Leguillon, D. Strength or toughness? A criterion for crack onset at a notch. *Eur. J. Mech. A/Solids* **2002**, *21*, 61–72. [CrossRef]
6. Carpinteri, A.; Cornetti, P.; Pugno, N.; Sapora, A.; Taylor, D. Generalized fracture toughness for specimens with re-entrant corners: Experiments vs. theoretical predictions. *Struct. Eng. Mech.* **2009**, *32*, 609–620. [CrossRef]
7. Sapora, A.; Firrao, D. Finite fracture mechanics predictions on the apparent fracture toughness of as-quenched Charpy V-type AISI 4340 steel specimens. *Fatigue Fract. Eng. Mater. Struct.* **2017**, *40*, 949–958. [CrossRef]
8. Susmel, L.; Taylor, D. On the use of the theory of critical distances to predict static failures in ductile metallic materials containing different geometrical features. *Eng. Fract. Mech.* **2008**, *75*, 4410–4421. [CrossRef]
9. Madrazo, V.; Cicero, S.; Carrascal, I.A. On the Point Method and the Line Method notch effect predictions in Al7075-T651. *Eng. Fract. Mech.* **2012**, *79*, 363–379. [CrossRef]
10. Cicero, S.; Fuentes, J.D.; Procopio, I.; Madrazo, V.; González, P. Critical distance default values for structural steels and a simple formulation to estimate the apparent fracture toughness in U-notched conditions. *Metals* **2018**, *8*, 871. [CrossRef]
11. Torabi, A.R.; Alaei, M. Mixed-mode ductile failure analysis of V-notched Al 7075-T6 thin sheets. *Eng. Fract. Mech.* **2015**, *150*, 70–95. [CrossRef]
12. Torabi, A.R.; Habibi, R.; Hosseini, B.M. On the ability of the Equivalent Material Concept in predicting ductile failure of U-notches under moderate- and large-scale yielding conditions. *Phys. Mesomech.* **2015**, *18*, 337–347. [CrossRef]
13. Torabi, A.R.; Alaei, M. Application of the equivalent material concept to ductile failure prediction of blunt V-notches encountering moderate-scale yielding. *Int. J. Damage Mech.* **2016**, *25*, 853–877. [CrossRef]
14. Torabi, A.R.; Keshavarzian, M. Tensile crack initiation from a blunt V-notch border in ductile plates in the presence of large plasticity at the notch vicinity. *Int. J. Terraspace Sci. Eng.* **2016**, *8*, 93–101.
15. Torabi, A.R. Estimation of tensile load-bearing capacity of ductile metallic materials weakened by a V-notch: The equivalent material concept. *Mater. Sci. Eng. A* **2012**, *536*, 249–255. [CrossRef]
16. Torabi, A.R.; Berto, F.; Campagnolo, A. Elastic-plastic fracture analysis of notched Al 7075-T6 plates by means of the local energy combined with the equivalent material concept. *Phys. Mesomech.* **2016**, *19*, 204–214. [CrossRef]
17. Fuentes, J.D.; Cicero, S.; Berto, F.; Torabi, A.R.; Madrazo, V.; Azizi, P. Estimation of Fracture Loads in AL7075-T651 Notched Specimens Using the Equivalent Material Concept Combined with the Strain Energy Density Criterion and with the Theory of Critical Distances. *Metals* **2018**, *8*, 87. [CrossRef]
18. Muñoz-Reja, M.; Távara, L.; Mantič, V.; Cornetti, P. Crack onset and propagation at fibre–matrix elastic interfaces under biaxial loading using finite fracture mechanics. *Compos. Part. A* **2016**, *82*, 267–278. [CrossRef]
19. Cornetti, P.; Sapora, A.; Carpinteri, A. Short cracks and V-notches: Finite Fracture Mechanics vs. Cohesive Crack Model. *Eng. Fract. Mech.* **2016**, *168*, 2–12. [CrossRef]
20. Cornetti, P.; Muñoz-Reja, M.; Sapora, A.; Carpinteri, A. Finite fracture mechanics and cohesive crack model: Weight functions vs. cohesive laws. *Int. J. Solid Struct.* **2019**, *156–157*, 126–136. [CrossRef]

21. Doitrand, A.; Estevez, R.; Leguillon, D. Comparison between cohesive zone and coupled criterion modeling of crack initiation in rhombus hole specimens under quasi-static compression. *Theor. Appl. Fract. Mech.* **2019**, *99*, 51–59. [CrossRef]

22. Felger, J.; Stein, N.; Becker, W. Mixed-mode fracture in open-hole composite plates of finite-width: An asymptotic coupled stress and energy approach. *Int. J. Solid. Struct.* **2017**, *122–123*, 14–24. [CrossRef]

23. Sapora, A.; Cornetti, P. Crack onset and propagation stability from a circular hole under biaxial loading. *Int. J. Fract.* **2019**, *214*, 97–104. [CrossRef]

24. Doitrand, A.; Leguillon, D. 3D application of the coupled criterion to crack initiation prediction in epoxy/aluminum specimens under four point bending. *Int. J. Solids Struct.* **2018**, *143*, 175–182. [CrossRef]

25. Stein, N.; Weißgraeber, P.; Becker, W. A model for brittle failure in adhesive lap joints of arbitrary joint configuration. *Compos. Struct.* **2015**, *133*, 707–718. [CrossRef]

26. Felger, J.; Rosendahl, P.L.; Leguillon, D.; Becker, W. Predicting crack patterns at bi-material junctions: A coupled stress and energy approach. *Int. J. Solids Struct.* **2019**, *164*, 191–201. [CrossRef]

27. Sapora, A.; Cornetti, P.; Carpinteri, A. Cracks at rounded V-notch tips: An analytical expression for the stress intensity factor. *Int. J. Fract.* **2014**, *187*, 285–291. [CrossRef]

28. Carpinteri, A.; Cornetti, P.; Sapora, A. Brittle failures at rounded V-notches: A finite fracture mechanics approach. *Int. J. Fract.* **2011**, *172*, 1–8. [CrossRef]

29. Filippi, S.; Lazzarin, P.; Tovo, R. Developments of some explicit formulas useful to describe elastic stress fields ahead of notches in plates. *Int. J. Solid. Struct.* **2002**, *39*, 4543–4565. [CrossRef]

Article

Critical Distance Default Values for Structural Steels and a Simple Formulation to Estimate the Apparent Fracture Toughness in U-Notched Conditions

Sergio Cicero [1], **Juan Diego Fuentes** [1,*], **Isabela Procopio** [1], **Virginia Madrazo** [2] and **Pablo González** [1]

[1] LADICIM (Laboratory of Materials Science and Engineering), University of Cantabria, E.T.S. de Ingenieros de Caminos, Canales y Puertos, Av/Los Castros 44, 39005 Santander, Spain; ciceros@unican.es (S.C.); pessoai@unican.es (I.P.); glezpablo@unican.es (P.G.)

[2] Centro Tecnológico de Componentes-CTC, C/Isabel Torres no. 1, 39011 Santander, Spain; madrazo.virginia@external.ensa.es

* Correspondence: fuentesjd@unican.es; Tel.: +34-942-200-928

Received: 11 October 2018; Accepted: 22 October 2018; Published: 24 October 2018

Abstract: The structural integrity assessment of components containing notch-type defects has been the subject of extensive research in the last few decades. The assumption that notches behave as cracks is generally too conservative, making it necessary to develop assessment methodologies that consider the specific nature of notches, providing accurate safe predictions of failure loads or defect sizes. Among the different theories or models that have been developed to address this issue the Theory of Critical Distances (TCD) is one of the most widely applied and extended. This theory is actually a group of methodologies that have in common the use of the material toughness and a length parameter that depends on the material (the critical distance; L). This length parameter requires calibration in those situations where there is a certain non-linear behavior on the micro or the macro scale. This calibration process constitutes the main practical barrier for an extensive use of the TCD in structural steels. The main purpose of this paper is to provide, through a set of proposed default values, a simple methodology to accurately estimate both the critical distance of structural steels and the corresponding apparent fracture toughness predictions derived from the TCD.

Keywords: fracture; critical distance; structural steel; notch

---

1. Introduction

There are numerous situations where the defects responsible for structural failure are not cracks (i.e., sharp defects whose tip radius tends to zero). If defects are blunt (e.g., notches), it may be overly conservative to assume that they behave like sharp cracks and, thus, to apply sharp crack analysis methodologies generally based on Fracture Mechanics. The reason is that notched components develop a load-bearing capacity that is greater than that developed by cracked components.

For brittle failure situations in cracked components, in which linear-elastic behavior is dominant, fracture mechanics establishes that fracture occurs when the applied stress intensity factor (K) is equal to the material fracture toughness (K_{mat}):

$$K = K_{mat} \tag{1}$$

Nevertheless, notches subject components to less severe stress fields at the defect tip, resulting in an apparent higher material fracture resistance (often referred to as apparent fracture toughness). If this is not taken into account in the analysis, Equation (1) often proves to be overly conservative.

Consequently, the specific nature of notches, and their consequences on the material behavior, has required the development of specific approaches for the fracture analysis of materials containing this type of defect. In this sense, the analysis of the fracture behavior of notches can already be performed using different criteria: The Theory of Critical Distances (TCD) [1–3], the Global Criterion [4,5], Process Zone models (e.g., [6,7]), statistical models (e.g., [8,9]), mechanistic models (e.g., [10]), the Strain Energy Density (SED) criterion (e.g., [11,12]), etc. Some of them are related to each other, so it is not straightforward to establish the boundaries between them. In any case, the TCD has been successfully applied to different failure mechanisms (e.g., fracture, fatigue) and materials, and is particularly simple to implement in structural integrity assessments (e.g., [1–3]). For these reasons, this work focuses on this particular approach.

2. The Theory of Critical Distances

The Theory of Critical Distances (TCD) comprises a set of methodologies with the common characteristic that they all use a characteristic material length parameter (the critical distance) when performing fracture assessments [1]. The origins of the TCD date back to the middle of the twentieth century [13,14], but it has been in the last two decades that this theory has seen an extensive development, providing answers to different scientific and engineering problems (e.g., [1–3,15–17]).

The above-mentioned length parameter is generally referred to as the critical distance, L, and in fracture analyses it follows Equation (2) [1]:

$$L = \frac{1}{\pi}\left(\frac{K_{\mathrm{mat}}}{\sigma_0}\right)^2 \tag{2}$$

where K_{mat} is the material fracture toughness (derived in cracked conditions) and σ_0 is a characteristic strength parameter, known as the inherent strength, which is generally higher than the ultimate tensile strength (σ_u) and requires calibration. When the material behavior is fully linear-elastic, σ_0 is equal to σ_u, and obtaining L is straightforward once the material fracture toughness and ultimate tensile strength are known.

Among the methodologies that comprise the TCD, the Point Method (PM) and the Line Method (LM) stand out both for their simplicity and their applicability.

Of these, the Point Method is the simplest methodology, and it affirms that fracture takes place when the stress at a distance of $L/2$ from the notch tip is equal to the inherent strength [1]. The resultant fracture criterion is, therefore:

$$\sigma\left(\frac{L}{2}\right) = \sigma_0 \tag{3}$$

The Line Method, meanwhile, assumes that fracture occurs when the average stress along a certain distance, $2L$, reaches the inherent strength [1]. Consequently, the LM follows:

$$\frac{1}{2L}\int_0^{2L}\sigma(r)dr = \sigma_0 \tag{4}$$

Furthermore, both the PM and the LM provide expressions for the apparent fracture toughness (K_{mat}^N) exhibited by materials containing U-notches when combined with the linear-elastic stress distribution at the notch tip provided by Creager and Paris [18]. The latter is equal to that ahead of the crack tip, but displaced a distance equal to $\rho/2$ along the x-axis (which is located in the notch midplane and has its origin at the crack tip [18]):

$$\sigma(r) = \frac{K}{\sqrt{\pi}}\frac{2(r+\rho)}{(2r+\rho)^{\frac{3}{2}}} \tag{5}$$

where K is the stress intensity factor for a crack with the same size as the notch, ρ is the notch radius, and r is the distance from the notch tip to the point being assessed. In order to keep in mind the validity range of the equations derived below, it should be noted that Equation (5) was derived for long thin notches (i.e., notch depth >> notch radius) and is only valid for small distances from the notch tip ($r <<$ notch depth).

If the PM is applied, Equation (3) can be combined with Equation (5), providing [1]:

$$K_{\mathrm{mat}}^N = K_{\mathrm{mat}} \frac{\left(1 + \frac{\rho}{L}\right)^{\frac{3}{2}}}{\left(1 + \frac{2\rho}{L}\right)} \tag{6}$$

This equation allows the apparent fracture toughness (K_{mat}^N) of a given material containing a U-shaped notch to be estimated from the material fracture toughness (K_{mat}, derived in cracked conditions), the notch radius (ρ), and the material critical distance (L). Analogously, when considering the LM (Equation (4)), together with Creager-Paris stress distribution (Equation (5)), the result is an even simpler equation [1]:

$$K_{\mathrm{mat}}^N = K_{\mathrm{mat}} \sqrt{1 + \frac{\rho}{4L}} \tag{7}$$

These equations have implications from a practical point of view, given that with any of them the fracture analysis of a notched component is reduced to an equivalent situation of a cracked component, with only the particularity of considering K_{mat}^N instead of K_{mat}. Accordingly, fracture takes place when:

$$K_I = K_{\mathrm{mat}}^N \tag{8}$$

Moreover, Equations (6) and (7) provide similar K_{mat}^N predictions. For this reason, the analysis shown below is focused on the LM predictions of K_{mat}^N (Equation (7)), although similar results would be derived from the PM (Equation (6)).

In addition, it should be noted that the authors have demonstrated [2,19] that notches may be analyzed by using Failure Assessment Diagrams (FADs) [20] and substituting K_{mat} with K_{mat}^N in the definition of the K_r coordinate of the assessment point. This coordinate is defined as the ratio between the applied stress intensity factor (K_I) and the material fracture resistance (K_{mat} for cracks and K_{mat}^N for notches) [20–22]. This work is not focused on the FAD approach, but the application of the K_{mat}^N values generated here to FAD analyses would be straightforward.

3. Materials and Methods

The authors have published a number of papers showing the application of the TCD to a wide range of structural steels: S275JR, S355J2, S460M and S690Q (e.g., [22,23]). These steels were tested at 3 different temperatures of their corresponding Ductile-to-Brittle Transition Zone (DBTZ) and, in the case of steels S275JR and S355J2, at temperatures equal to their Lower Shelf (LS). Hence, the resultant experimental program collected here includes 15 different mechanical behaviors, which are summarized in Table 1 [23,24]. Notch radii varied between 0 mm (crack-like defects) up to 2.0 mm in all cases.

Table 1. Summary of the experimental results analyzed in this paper. LS: Lower Shelf; DBTZ: Ductile-to-Brittle Transition Zone [23,24].

Steel	Number of Tests	K_{mat} (MPa·m$^{1/2}$)	L (mm)	σ_u (MPa)	σ_0 (MPa)
S275JR (−120 °C, LS)	23	48.80	0.0137	614	7438
S275JR (−90 °C, LS)	24	62.72	0.0062	597	14,211
S275JR (−50 °C, DBTZ)	24	80.60	0.0049	565	20,543
S275JR (−30 °C, DBTZ)	24	100.70	0.0061	549	23,003
S275JR (−10 °C, DBTZ)	34	122.80	0.0083	536	24,048
S355J2 (−196 °C, LS)	24	31.27	0.0198	923	3965
S355J2 (−150 °C, DBTZ)	21	60.56	0.0084	758	11,789
S355J2 (−120 °C, DBTZ)	22	146.60	0.0168	672	20,179
S355J2 (−100 °C, DBTZ)	35	157.40	0.0140	647	23,734
S460M (−140 °C, DBTZ)	24	45.60	0.0028	795	15,375
S460M (−120 °C, DBTZ)	24	88.29	0.0075	759	18,189
S460M (−100 °C, DBTZ)	33	88.58	0.0053	727	21,708
S690Q (−140 °C, DBTZ)	24	69.11	0.0069	1112	14,844
S690Q (−120 °C, DBTZ)	24	103.80	0.0131	1061	16,180
S690Q (−100 °C, DBTZ)	34	125.40	0.0170	1016	17,159

The fracture toughness tests (in cracked specimens) and the apparent fracture toughness tests (in notched specimens) were performed following ASTM1820 standard [25]; alternative experimental approaches can be found in literature (e.g., [26–30]), whereas the L values of the four steels at the different temperatures (see Table 1) were calibrated by a least squares fitting of the experimental results. Figure 1 [23,24] shows an example corresponding to steel S275JR tested at −120 °C. σ_0 values were directly obtained from Equation (2) once K_{mat} and L were known. The total number of tests is 394, with L values varying from 0.0028 mm up to 0.0198 mm. Thus, the experimental results collected here represent a wide range of situations.

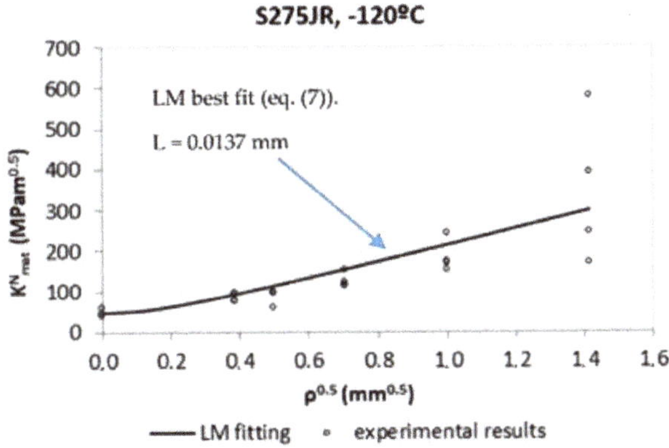

Figure 1. Experimental notch fracture toughness results, Line Method (LM) fitting and derivation of the corresponding critical distance. Steel S275JR at −120 °C.

4. Derivation of Default Values of *L* and Apparent Fracture Toughness Estimations

Once the experimental (best fit calibrated) values of L have been defined, and the corresponding values of σ_0 have been derived (through Equation (2)), and provided that σ_u values are known for each material condition, it is proposed to establish the (non-dimensional) relation between the inherent strength (σ_0) and the ultimate tensile strength (σ_u):

$$m = \frac{\sigma_0}{\sigma_u} \tag{9}$$

For each structural steel and testing temperature a conservative value of m may be obtained. This value of m, combined with the ultimate tensile strength of the material (σ_u), may substitute the inherent strength (σ_0) in Equation (2), resulting in a default (conservative) critical distance (L_d) that does not need to be calibrated and allows the TCD to be applied safely:

$$L_d = \frac{1}{\pi}\left(\frac{K_{mat}}{m \cdot \sigma_u}\right)^2 \tag{10}$$

This methodology was presented by the authors [31], providing default conservative values for different types of materials: steels, aluminum alloys, polymers, ceramic and rocks, and composites. This work was based on a vast database of experimental results obtained on notched fracture specimens. The present work, however, particularizes the analysis to structural steels, providing more accurate (yet conservative) default values of L for structural steels.

Figures 2 and 3 show the experimental value of m for each material analyzed. It can be observed that Figure 2 includes additional data taken from the literature [32,33]. The values have to be analyzed separately in two different groups, attending to their different fracture behavior and mechanical properties: structural steels operating in LS conditions, and structural steels operating in the DBTZ.

In the LS regime, with the structural steels having brittle behavior, the resulting lower bound value of m gathered in [31] is 1.6, as can be seen in Figure 2. However, some of the values may be much higher, close to 25, although most of them are below 5.0. For this reason, a fitting curve has been developed with the aim of improving the accuracy of the results (when compared to those provided by a single value of m), yet providing conservative values (σ_u in MPa):

$$m = \frac{1.3 \cdot \sigma_u}{\sigma_u - 490} \tag{11}$$

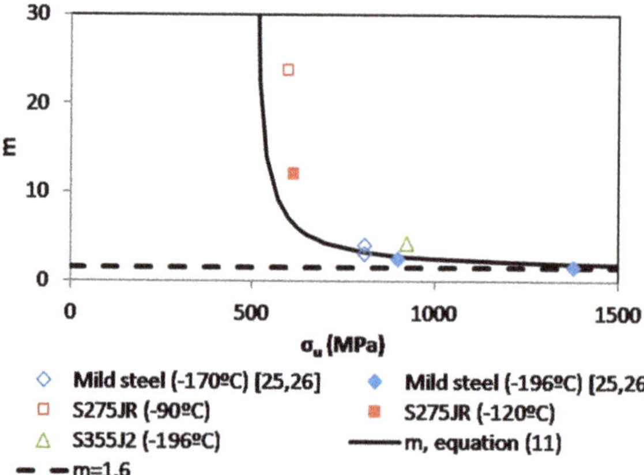

Figure 2. m values versus tensile strength for structural steels operating at the Lower Shelf.

Figure 2 reveals the limitations of the lower bound value proposed in [31]. Although it provides a safe estimate of m (and then, also of L and K_{mat}^N), it is still clearly over-conservative for structural steels with relatively low ultimate tensile strengths (e.g., 500–800 MPa), which is a very typical situation in practice. The benefits of using Equation (11) instead of the lower bound value proposed in [31] are, thus, evident.

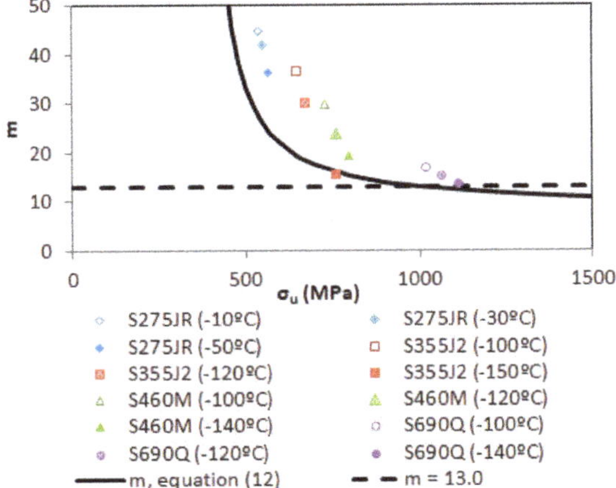

Figure 3. *m* values versus tensile strength for structural steels operating at the Ductile-to-Brittle Transition Zone. Derivation of a safe estimation (*m* = 13.0).

For structural steels operating in the DBTZ the lower bound value of *m* proposed in Reference [31] was 13, as shown in Figure 3. This much higher correction is associated with a more non-linear behavior of the structural steels within the DBTZ when compared to the behavior of the same steels within the LS. Again, a fitting curve is provided with the aim of improving the accuracy of the predictions (σ_u in MPa):

$$m = \frac{8 \cdot \sigma_u}{\sigma_u - 380} \tag{12}$$

When comparing the lower bound value proposed in [31] (*m* = 13) with Equation (12), the conclusions are analogous to those derived in LS situations. The lower bound value looks accurate for high ultimate tensile strengths (e.g., higher than 1000 MPa), but it is clearly over-conservative for steels with ultimate tensile strength ranges between approximately 500 MPa and 800 MPa.

Here, it should be noted that the validity range of all these proposed *m* values is limited to the steel grades covered by this work. With all this, the LM K_{mat}^N estimations can be easily derived through the following equation:

$$K_{mat}^N = K_{mat}\sqrt{1 + \frac{\rho}{4L_d}} \tag{13}$$

Figures 4–7 show the predictions for the 394 tests. It can be observed that, in order to perform a homogeneous analysis, both the apparent fracture toughness values and the notch radii values have been normalized by the fracture toughness obtained in cracked conditions and by the corresponding critical distance, in a (K_{mat}^N / K_{mat}) against (ρ/L_d)$^{1/2}$ plot.

The two lines on each figure correspond to two different values of K_{mat} considered in the analysis to be introduced in Equation (13): the average value of the experimental results (K_{mat}) for each particular steel and testing temperature, and $K_{mat,0.95}$, which is associated to a 95% confidence level. The latter has been derived assuming a normal distribution of the experimental results obtained in cracked conditions (K_{mat}), and would be the prediction required for structural integrity assessment purposes, whereas the former is the prediction that should better capture the physics of the phenomenon being analyzed. The above referred confidence level is, therefore, limited for the fracture results obtained in cracked conditions ($\rho = 0$ mm). For notched conditions (i.e., for the rest of the curve), using $K_{mat,0.95}$ provides a more conservative estimation of the apparent fracture toughness than that obtained with the average value of the fracture toughness (K_{mat}), but the corresponding predictions are not necessarily associated to a 95% confidence level.

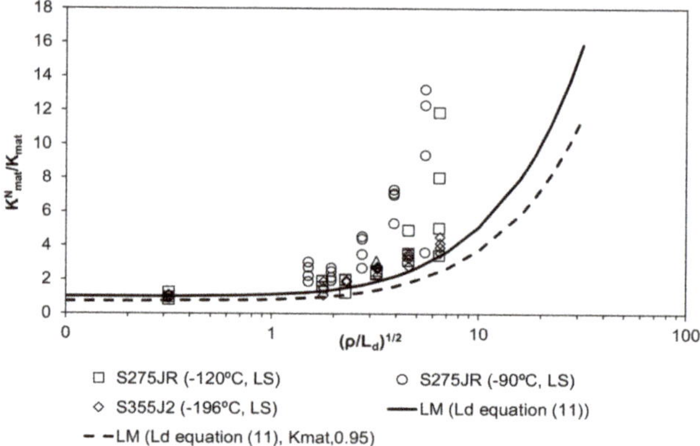

Figure 4. Apparent fracture toughness predictions (*m* derived from Equation (11)) and comparison with the experimental results. Structural steels operating at Lower Shelf temperatures.

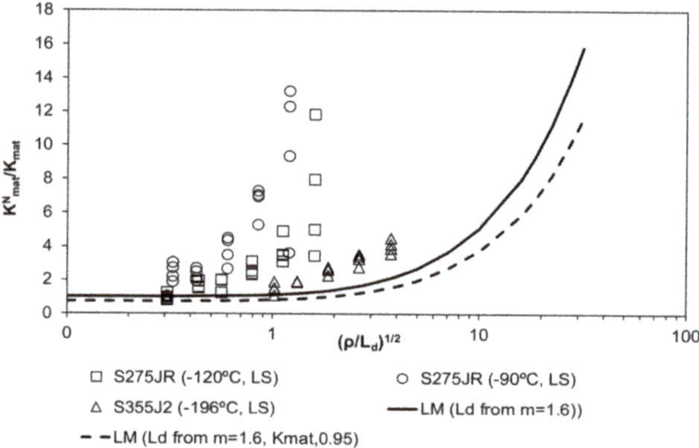

Figure 5. Apparent fracture toughness predictions (*m* = 1.6) and comparison with the experimental results. Structural steels operating at Lower Shelf temperatures.

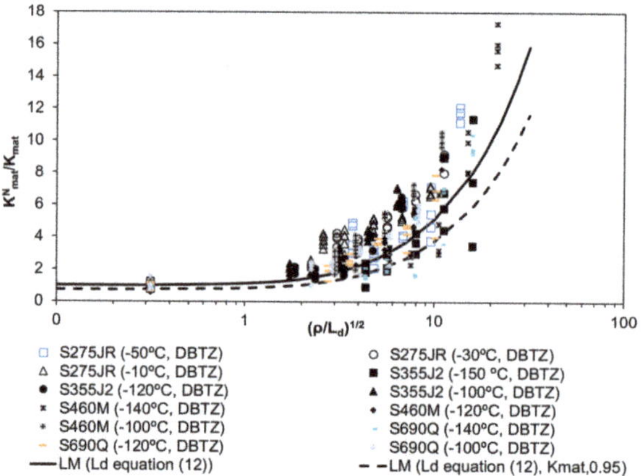

Figure 6. Apparent fracture toughness predictions (*m* derived from Equation (12)) and comparison with the experimental results. Structural steels operating at the DBTZ.

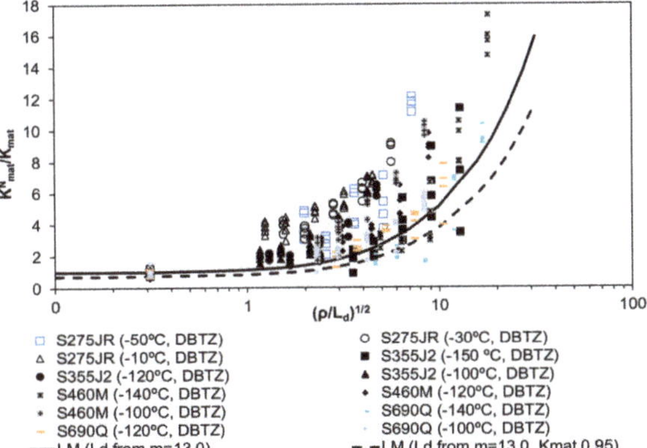

Figure 7. Apparent fracture toughness predictions ($m = 13.0$) and comparison with the experimental results. Structural steels operating at the DBTZ.

From the results shown in Figures 4–7, the following observations can be made:

1. The LM predictions derived from the proposed default values of the material critical distance (L_d) capture a significant part of the physics of the notch effect, given that the LM prediction adequately follows the tendency of the experimental results, which have been obtained for a wide variety of structural steels and conditions. The results are particularly accurate considering fitting Equations (11) and (12) and the average value of the material fracture toughness (K_{mat}) for each particular steel and working temperature. These fitting equations significantly reduce the conservatism obtained from the lower bound values proposed in [31].

2. The most conservative results have been obtained in Lower Shelf conditions when using the lower bound value of m ($m = 1.6$). Even under such circumstances, the resulting apparent fracture toughness estimations may be significantly higher than the corresponding fracture toughness obtained in cracked conditions, so the potential reduction of conservatism is still important.

3. If the LM evaluations are to be used in structural integrity assessments, although the use of K_{mat} (average value of the material fracture toughness obtained in cracked conditions) captures most of the notch effect adequately, it may be unsafe on many occasions due to the high scatter of the fracture processes. This means that it sometimes provides apparent fracture toughness values higher than those measured experimentally (see Figures 4 and 6). In order to provide a fracture analysis tool to be used in structural integrity assessments, it is necessary to propose a methodology that is capable of providing safe predictions of the apparent fracture toughness. With this purpose, it is proposed to use a 95% confidence level value of the fracture toughness ($K_{mat,0.95}$).

4. Thus, the most accurate, yet conservative, methodology for the apparent fracture toughness estimation arises from the combination of $K_{mat,0.95}$ (as the material fracture toughness) and the m values derived from Equations (11) and (12), for LS and DBTZ conditions respectively.

5. Summary

The aim of this paper has been to provide a simple accurate methodology to estimate both the critical distance of structural steels and the corresponding apparent fracture toughness predictions derived from the Theory of Critical Distances, and, particularly, from the Line Method. This has been done through a set of proposed factors (m values) that multiply the material ultimate tensile strength, avoid any need for previous calibration, and allow default values of the critical distance (L_d) to be obtained. The m factors have been provided as fitting curves depending exclusively on the material ultimate tensile strength. Results are also shown when using previously proposed m lower bound values, which provide more conservative estimations of the material apparent fracture toughness.

The methodology has been applied to four structural steels (S275JR, S355J2, S460M, and S690Q) tested on notched conditions (U-shaped notches) and operating at both the Lower Shelf and the Ductile-to-Brittle Transition Zone. The experimental values of the apparent fracture toughness (K_{mat}^N) and the notch radii values have been normalized by the corresponding fracture toughness (K_{mat} or $K_{mat,0.95}$) obtained in cracked conditions and by the default values derived for the critical distance (L_d) respectively, representing the 394 tests in (K_{mat}^N/K_{mat}) − $(\rho/L_d)^{1/2}$ plots.

The results demonstrate the capacity of the proposed methodology to provide safe estimations of the apparent fracture toughness whilst capturing a significant part of the notch effect and, thus, reducing the conservatism associated with the assumption that notches behave as cracks.

Author Contributions: S.C. and V.M. conceived and designed the experiments; V.M., J.D.F. and I.P. performed the experiments; all the authors analyzed the data; S.C. wrote the paper.

Funding: This research received no external funding.

Acknowledgments: The authors of this work would like to express their gratitude to the Spanish Ministry of Science and Innovation for the financial support of the Project MAT2014-58443-P: "Análisis del comportamiento en fractura de componentes estructurales con defectos en condiciones de bajo confinamiento tensional", on the results of which this paper is based.

References

1. Taylor, D. *The Theory of Critical Distances: A New Perspective in Fracture Mechanics*; Elsevier: Oxford, UK, 2007; ISBN 978-008044478-9.
2. Cicero, S.; Madrazo, V.; Carrascal, I.A.; Cicero, R. Assessment of notched structural components using failure assessment diagrams and the theory of critical distances. *Eng. Fract. Mech.* **2011**, *78*, 2809–2825. [CrossRef]
3. Cicero, S.; Madrazo, V.; Carrascal, I.A. Analysis of notch effect in PMMA using the theory of critical distances. *Eng. Fract. Mech.* **2012**, *86*, 56–72. [CrossRef]
4. Niu, L.S.; Chehimi, C.; Pluvinage, G. Stress field near a large blunted V notch and application of the concept of notch stress intensity factor to the fracture of very brittle materials. *Eng. Fract. Mech.* **1994**, *49*, 325–335.
5. Pluvinage, G. Fatigue and fracture emanating from notch; the use of the notch stress intensity factor. *Nucl. Eng. Des.* **1998**, *185*, 173–184. [CrossRef]
6. Dugdale, D.S. Yielding of steel sheets containing slits. *J. Mech. Phys. Solids* **1960**, *8*, 100–108. [CrossRef]
7. Gómez, F.J.; Elices, M.; Valiente, A. Cracking in PMMA containing U-shaped notches. *Fat. Frac. Eng. Mat. Struct.* **2000**, *23*, 795–803. [CrossRef]
8. Weibull, W. The phenomenon of rupture in solids. *Proc. R. Swed. Inst. Eng. Res.* **1939**, *153*, 1–55.
9. Beremin, F.M.; Pineau, A.; Mudry, F.; Devaux, J.C.; D'Escatha, Y.; Ledermann, P. A local criterion for cleavage fracture of a nuclear pressure vessel steel. *Metall. Trans. A* **1983**, *14A*, 2277–2287. [CrossRef]
10. Ritchie, R.O.; Knott, J.F.; Rice, J.R. On the relationship between critical tensile stress and fracture toughness in mild steel. *J. Mech. Phys. Solids* **1973**, *21*, 395–410. [CrossRef]
11. Sih, G.C. Strain-energy-density factor applied to mixed mode crack problems. *Int. J. Fract.* **1974**, *10*, 305–321. [CrossRef]
12. Berto, F.; Lazzarin, P. Recent developments in brittle and quasi-brittle failure assessment of engineering materials by means of local approaches. *Mater. Sci. Eng. R* **2014**, *75*, 1–48. [CrossRef]
13. Neuber, H. *Theory of Notch Stresses: Principles for Exact Calculation of Strength with Reference to Structural form and Material*; Springer: Berlin, Germany, 1958.
14. Peterson, R.E. Notch sensitivity. In *Metal Fatigue*; McGraw Hill: New York, NY, USA, 1959; pp. 293–306.
15. Susmel, L.; Taylor, D. On the use of the Theory of Critical Distances to predict failures in ductile metallic materials containing different geometrical features. *Eng. Fract. Mech.* **2008**, *75*, 4410–4421. [CrossRef]
16. Susmel, L.; Taylor, D. An elasto-plastic reformulation of the Theory of Critical Distances to estimate lifetime of notched components failing in the low/medium-cycle fatigue regime. *J. Eng. Mater. Technol.* **2010**, *132*, 0210021–0210028. [CrossRef]

17. Taylor, D. A mechanistic approach to critical-distance methods in notch fatigue. *Fatig. Fract. Eng. Mater. Struct.* **2001**, *24*, 215–224. [CrossRef]

18. Creager, M.; Paris, P.C. Elastic field equations for blunt cracks with reference to stress corrosion cracking. *Int. J. Fract.* **1967**, *3*, 247–252. [CrossRef]

19. Madrazo, V.; Cicero, S.; García, T. Assessment of notched structural steel components using failure assessment diagrams and the theory of critical distances. *Eng. Fail. Anal.* **2014**, *36*, 104–120. [CrossRef]

20. Anderson, T.L. *Fracture Mechanics: Fundamentals and Applications*, 3rd ed.; CRC Press: Boca Raton, FL, USA, 2005.

21. BS 7910. *Guide to Methods for Assessing the Acceptability of Flaws in Metallic Structures*; British Standards Institution: London, UK, 2013.

22. Kocak, M.; Webster, S.; Janosch, J.J.; Ainsworth, R.A.; Koers, R. (Eds.) *FITNET Fitness-for-Service (FFS) Procedure—Volume 1*; GKSS Forschungzscentrum: Geesthacht, Germany, 2008.

23. Cicero, S.; Madrazo, V.; García, T. Analysis of notch effect in the apparent fracture toughness and the fracture micromechanisms of ferritic–pearlitic steels operating within their lower shelf. *Eng. Fail. Anal.* **2014**, *36*, 322–342. [CrossRef]

24. Cicero, S.; García, T.; Madrazo, V. Application and validation of the Notch Master Curve in medium and high strength structural steels. *J. Mech. Sci. Tech.* **2015**, *29*, 4129–4142. [CrossRef]

25. ASTM E1820-09e1. *Standard Test Method for Measurement of Fracture Toughness*; American Society for Testing and Materials: Philadelphia, PA, USA, 2009.

26. Quinn, G.D.; Bradt, R.C. On the Vickers Indentation Fracture Test. *J. Am. Ceram. Soc.* **2007**, *90*, 673–680. [CrossRef]

27. Akono, A.T.; Randall, N.X.; Ulm, F.J. Experimental determination of the fracture toughness via microscratch tests: Application to polymers, ceramics, and metals. *J. Mater. Res.* **2012**, *27*, 485–493. [CrossRef]

28. Akono, A.T.; Alm, F.J. An improved technique for characterizing the fracture toughness via scratch test experiments. *Wear* **2014**, *313*, 117–124. [CrossRef]

29. Sola, R.; Giovanardi, R.; Parigi, G.; Varonesi, P. A Novel Method for Fracture Toughness Evaluation of Tool Steels with Post-Tempering Cryogenic Treatment. *Metals* **2017**, *7*, 75. [CrossRef]

30. Lacalle, R.; Álvarez, J.A.; Gutiérrez-Solana, F. Use of Small Punch Notched Specimens in the Determination of Fracture Toughness. *ASME Press. Vessels Pip. Conf.* **2008**, *6*, 1363–1369. [CrossRef]

31. Fuentes, J.D.; Cicero, S.; Procopio, I. Some default values to estimate the critical distance and their effect on structural integrity assessments. *Theor. Appl. Fract. Mech.* **2017**, *90*, 204–212. [CrossRef]

32. Taylor, D.; Cornetti, P.; Pugno, N. The fracture mechanics of finite crack extension. *Eng. Fract. Mech.* **2005**, *72*, 1021–1038. [CrossRef]

33. Susmel, L.; Taylor, D. The Theory of Critical Distances as an alternative experimental strategy for the determination of KIC and ΔKth. *Eng. Fract. Mech.* **2010**, *77*, 1492–1501. [CrossRef]

Communication

Feasibility Study on Application of Synchrotron Radiation μCT Imaging to Alloy Steel for Non-Destructive Inspection of Inclusions

Yoshinobu Shimamura [1,*], Shinya Matsushita [2], Tomoyuki Fujii [1], Keiichiro Tohgo [1], Koichi Akita [3], Takahisa Shobu [4] and Ayumi Shiro [5]

[1]   Department of Mechanical Engineering, Shizuoka University, Shizuoka 432-8561, Japan; fujii.tomoyuki@shizuoka.ac.jp (T.F.); tohgo.keiichiro@shizuoka.ac.jp (K.T.)

[2]   Hitachi Construction Machinery Co., Ltd., Tokyo 110-0015, Japan; m.sny.108@gmail.com

[3]   Department of Mechanical Systems Engineering, Tokyo City University, Tokyo 158-8557, Japan; akitak@tcu.ac.jp

[4]   Japan Atomic Energy Agency, Hyogo 679-5148, Japan; shobu@sp8sun.spring8.or.jp

[5]   National Institutes for Quantum and Radiological Science and Technology, Hyogo 679-5148, Japan; shiro.ayumi@qst.go.jp

*   Correspondence: shimamura.yoshinobu@shizuoka.ac.jp; Tel.: +81-53-478-1045

Received: 15 April 2019; Accepted: 5 May 2019; Published: 8 May 2019

Abstract: In order to examine the feasibility of applying synchrotron radiation μCT imaging to alloy steels for non-destructive inspection of inclusions for potential origins of internal fatigue damage in the very high cycle region, synchrotron radiation μCT imaging was utilized for repeated non-destructive observation of Cr-Mo steel. An ultrasonic fatigue testing machine was used in aid of the repeated observation. As a result, it was found that the synchrotron radiation μCT imaging with 70 keV was useful for non-destructive observation of inclusions of more than 10 μm, one of which may be an internal fatigue origin. No identifiable damage was observed around every inclusion, and in the base metal, at least up to 70% of fatigue life was observed in the imaging volume.

Keywords: fatigue; X-ray techniques; alloy steel; synchrotron radiation; μCT imaging; internal fatigue fracture

---

1. Introduction

Alloy steels may suffer from a fatigue fracture called a "Giga cycle fatigue" or a "Very high cycle fatigue," which is a fatigue phenomenon as a result of cyclic loading in excess of 10^7 cycles. The fatigue fracture in the very high cycle region often originates from an inclusion such as metallic oxide. A small, dark area, whose diameter is a few 10 μm, is observed around the inclusion when optical microscopy is used, and a fine granular fracture surface was observed in the dark area if scanning electron microscopy is used. The circular area is referred to as a "fish eye" [1], and the dark area in the center of a fish eye is referred to as "optical dark area (ODA)" [2] or "Fine granular area (FGA)" [3]. In this study, we use a terminology "ODA" to indicate the area. The diameter of ODA is a few tens μm.

It has been suggested [4–6] that the nucleation and the formation of ODA consume most of the fatigue life for an internal fatigue fracture. Thus, it is important to elucidate the mechanism of nucleation and the formation process of ODA. Several experimental papers [4–6] have been reported for tackling with the problem. Kuroshima et al. [4] conducted elaborative experiments in which specimens subjected to cyclic loading were cut and polished until inclusions were found, and concluded that penny-shaped microcracks around many inclusions nucleated in the early stage of very high cycle fatigue of Cr-Mo steel, but the propagation of the microcracks was very slow and the size of the microcracks slightly increased just before fatigue failure. Lu and Shiozawa [5] and Ogawa et al. [6]

measured the crack growth rate of internal fatigue crack of high C-Cr spring steel by using a beach mark technique and concluded that the crack growth rate in ODA was extremely slow compared to that of surface cracks. For titanium alloys, internal fatigue fracture in the very high cycle region can be seen [7,8]. Yoshinaka [7] reported non-destructive observation of internal fatigue crack nucleation and propagation of Ti-6Al-4V by using synchrotron radiation μCT imaging. However, to the best of our knowledge, there are no results in the literature regarding non-destructive observation of nucleation and propagation of the ODA of ferrous alloys. Therefore, synchrotron radiation μCT imaging was used in this study for the non-destructive observation of Cr-Mo steel subjected to fatigue loading up to 10^8 cycles.

Synchrotron radiation μCT imaging is a micro computed tomography imaging technique that uses high energy X-ray of about 100 keV available in synchrotron radiation facilities, and a high spatial resolution on the order of μm can be achieved even if the target material is a heavy metal such as Fe [9]. In this study, SPring-8 in Hyogo, Japan was used, which is one of the largest third-generation synchrotron facilities. Since the available time of a beamline in SPring-8 was limited to within 48 h, an ultrasonic fatigue testing machine with the cyclic frequency of 20 kHz was used to exert fatigue loading on site, allowing us to conduct repeated μCT imaging of an identical specimen from an intact condition, i.e., before applying cyclic loading, to fatigue-loaded conditions up to 10^8 cycles within 48 h.

The aim and novelty of the present work are to examine the feasibility of non-destructive inspection of inclusions, one of which may be an internal fatigue origin site, and to examine the feasibility of detection of internal fatigue damage, i.e., fatigue crack nucleation and progress, in alloy steels using synchrotron radiation μCT imaging in aid of an ultrasonic fatigue testing technique for reducing fatigue testing periods on site. In addition, the fatigue strength was compared with the prediction using the Vickers hardness and inclusion size, the so-called Murakami's approach [10], to confirm the effect of the strain rate on the fatigue strength in this study.

2. Material and Methods

2.1. Material

Cr-Mo steel, JIS SCM420H, was used. The chemical composition is shown in Table 1. Specimens were annealed at 900 °C for 45 min, oil quenched, and tempered at 180 °C for 120 min. The proof stress was about 850 MPa, the tensile strength was about 1200 MPa, and the Vickers hardness was 435 Hv. The specimen's surface was mirror finished after heat treatment to minimize the influence of surface roughness on the quality of radiographs.

Table 1. Chemical composition of SCM420H (wt%).

C	Si	Mn	P	S	Cu	Cr	Ni	Mo	Fe
0.2	0.18	0.8	0.02	0.02	0.01	1.12	0.01	0.19	Bal.

2.2. Ultrasonic Fatigue Testing Machine

An ultrasonic axial fatigue testing machine is mainly composed of a piezoelectric oscillator, an amplifying horn, a specimen, and a controller, as shown in Figure 1 [8]. Sinusoidal electrical input from the signal generator vibrates the piezoelectric oscillator at about 20 kHz. The amplitude of the vibration is, however, only a few μm and not enough to bring fatigue failure to a specimen. Thus, the cyclic vibration of the piezoelectric oscillator is amplified using an amplifying horn. The specimen had the first natural frequency at about 20 kHz of axial vibration mode and was resonated by the external input from the amplifying horn. In this study, an hourglass type specimen was designed by using theoretical formulas [11], and then the specimen shape was tuned by using finite element analysis to obtain exact resonance. The specimen shape and the profiles of axial stress amplitude are presented in Figure 2. In this shape, the maximum axial stress amplitude was achieved at the center of the necked region of the specimen, whose diameter was 2.5 mm. The stress concentration factor K_t at the root of the necked portion was 1.03. The stress ratio R was −1. Fatigue tests were conducted in air with forced cooling using compressed air. Intermittent loading was used to keep the specimen temperature below 100 °C. A loading time of 0.3 s and a dwelling time of 0.4 s were set for the intermittent testing.

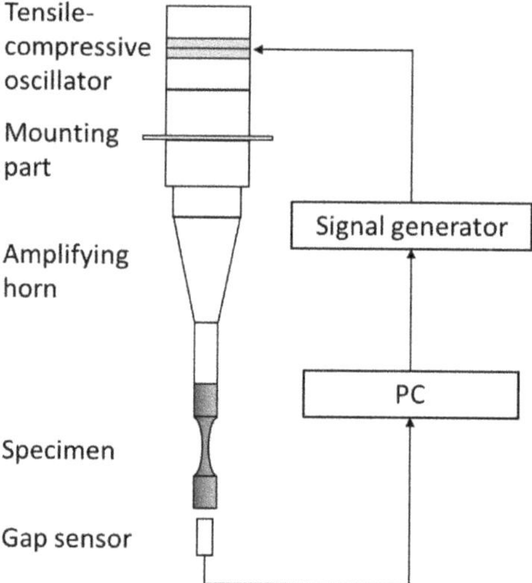

Figure 1. Configuration of an ultrasonic fatigue testing machine.

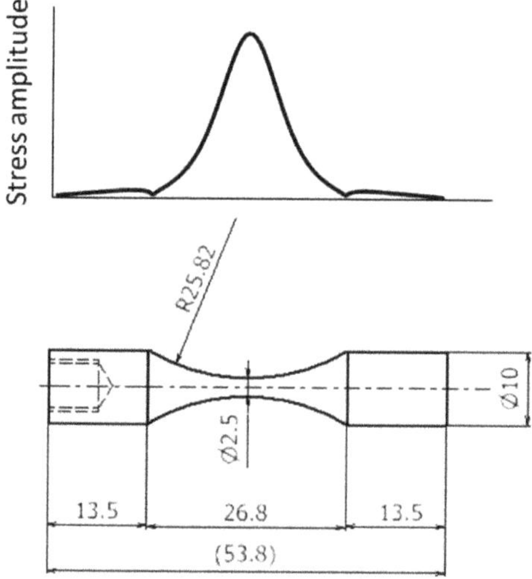

Figure 2. Specimen shape and the profile of the axial stress amplitude, along with the specimen axis.

The stress amplitude was estimated as follows. First, the relation between the axial displacement amplitude at the end of a specimen and the stress amplitude at the center in the necked region was calculated using finite element analysis. During fatigue testing, the axial displacement at the end of a specimen was monitored using a gap sensor, and the stress amplitude was estimated by substituting the measured axial displacement amplitude into the analytical relation.

To detect fatigue crack initiation, the resonance frequency was monitored during fatigue testing. If the resonance frequency decreased by 50 Hz, fatigue testing was interrupted.

Preliminary fatigue tests were conducted in order to decide the fatigue strength in the very high cycle region of the prepared specimens, and we decided to conduct fatigue testing at $\sigma_a = 650$ MPa for synchrotron radiation μCT imaging because it is anticipated that a specimen will fail in the range between 10^7 to 10^8 cycles by internal fracture from inclusion.

2.3. Synchrotron Radiation μCT Imaging

Synchrotron radiation μCT imaging was conducted at the beamline BL22XU at SPring-8. The maximum X-ray energy was about 70 keV. A specimen was placed on a rotating table so that the specimen axis was arranged on the rotating axis. The set-up of synchrotron radiation μCT imaging is shown in Figure 3. Radiographs were taken at intervals of 0.2 degrees from 0 degrees to 180 degrees; the total number of radiographs was about 900 images. The area of radiographs was equal to the area of incident X-ray, which was H2.1 mm × W4.0 mm, where the width was wider than the diameter of the necked portion of the specimen, i.e., 2.5 mm. The pixel size was 6.5 μm. The stress amplitude at the top and bottom edge of the imaging volume was 95% of the maximum applied stress amplitude of the measured area. During μCT imaging, a specimen was subjected to a static tensile load of 500 MPa, which was lower than the fatigue strength at 10^9 cycles, by using an in-house tensile loading apparatus in order to open if an internal crack(s) exists. The in-house tensile loading apparatus was mainly composed of an acrylic tube and a coil spring; a specimen was placed in the acrylic tube, fixed at the one end of the acrylic tube, and loaded with the coil spring by tightening a screw located in the other end. A preliminary μCT imaging using a specimen with a fatigue crack from the surface was conducted to assess the feasibility of crack detection. The radiograph of the crack tip implied that it seemed to be possible to identify an internal fatigue crack if the size is on the order of 10 μm.

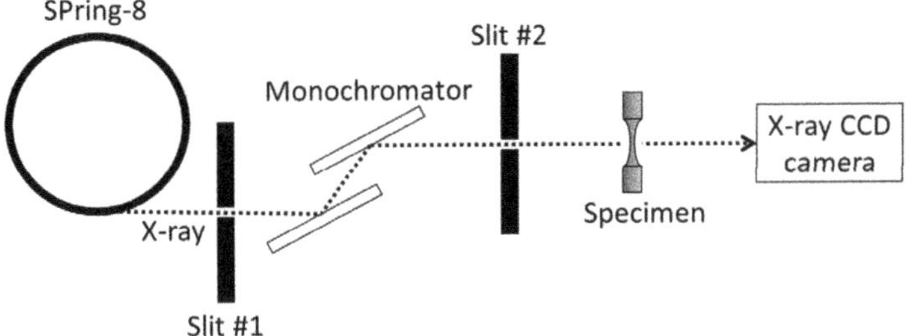

Figure 3. Set-up of the synchrotron radiation μCT imaging.

The first μCT imaging was conducted for a specimen before fatigue testing (N = 0). Then, several sets of fatigue loading were repeated by using an ultrasonic fatigue testing machine and μCT imaging. The cumulative numbers of cycles at μCT imaging were 0, 10^7, 2×10^7, and 10^8. Owing to the time limitation for μCT imaging at SPring-8, the repeated μCT imaging was conducted for only one specimen.

For reconstruction of radiographs, software provided by SPring-8 [12] was used.

2.4. Determination of Fatigue Life of the Specimen Used for the Repeated μCT Imaging

After the μCT imaging experiment at SPring-8, the specimen was subjected to fatigue loading until fatigue failure in our laboratory by using the same fatigue testing machine and conditions mentioned above. The specimen failed when the cumulative number of cycles reached 1.4×10^8 cycles from an internal inclusion as shown in Figure 4. The result means that repeated μCT imaging was conducted at 0%, 7.1%, 14.2%, and 70.9% of the fatigue life in the experiments. The chemical composition of the origin was characterized by using EPMA (JXA-8530F, JEOL, Tokyo, Japan) and was identified as an inclusion of Al and Ca oxide with a 30 μm diameter. It should be noted that the cracked position was 3.1 mm far from the center of the imaging volume, which was, unfortunately, out of the imaging volume.

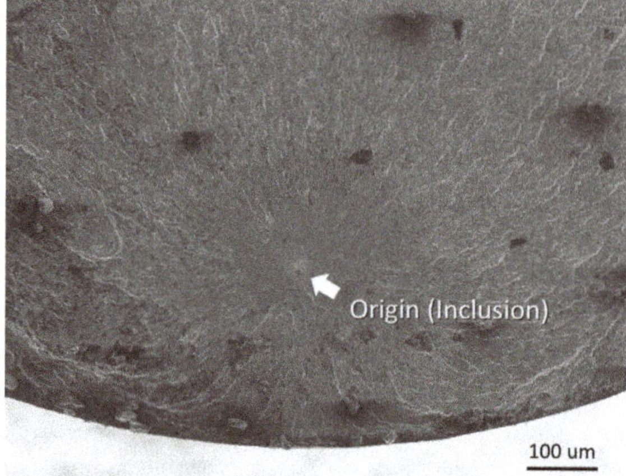

Figure 4. Fatigue origin of the specimen used for the repeated synchrotron radiation μCT imaging.

3. Results

An example of a radiograph is shown in Figure 5, and an example of the reconstructed image of a cross section is shown in Figure 6. In the reconstructed image, a block of black pixels is observed. The block of black pixels was considered to be an inclusion [9]. At least 47 inclusions were identified within the imaging volume. Figure 7 shows the locations of the inclusions identified from the reconstructed images before fatigue loading. The projected positions in the vertical and horizontal sections are shown in the left and right figure, respectively. Most of the block size varied from 2 to 6 pixels, which corresponds to 12 μm to 36 μm in diameter. Since an inclusion size as a fatigue origin is, in general, 10 μm or more, the μCT imaging technique used can identify inclusions that may be fatigue origins. In other words, 47 potential fatigue crack origins were non-destructively monitored using the μCT imaging technique in the experiment, even though the repeated μCT imaging was conducted for only one specimen due to time limitation.

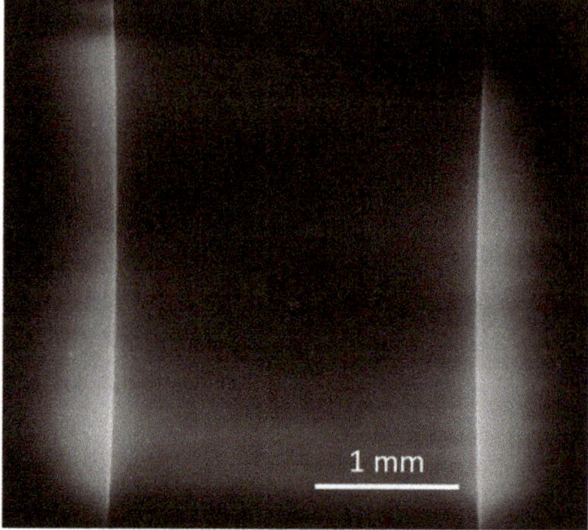

Figure 5. Example of a radiograph.

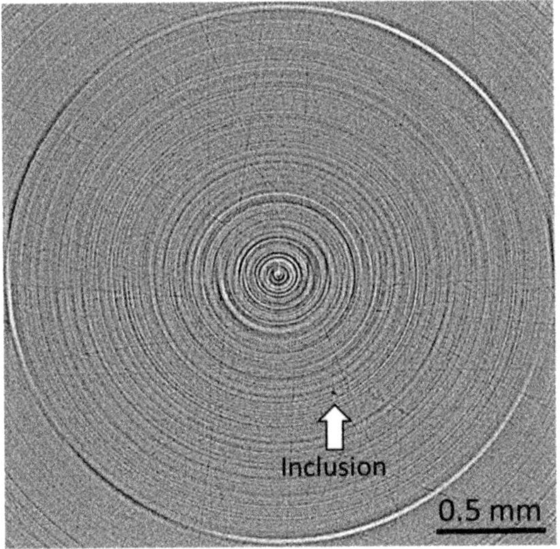

Figure 6. Example of the reconstructed image of a cross section.

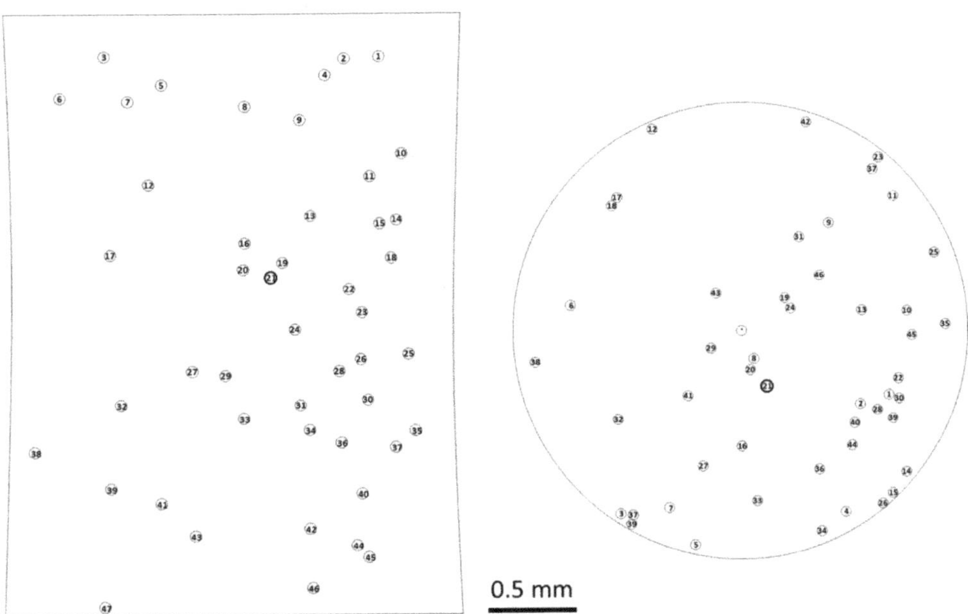

Figure 7. Locations of inclusions.

Then, reconstructed images before and after fatigue loading were compared for investigating crack formation from inclusions. An example of reconstructed images of an inclusion subjected to almost the maximum stress amplitude, which is labeled "21" and thick circled in Figure 7, is shown in Figure 8. If a fatigue crack exits around an inclusion, crack opening occurs due to the applied load exerted by the in-house tensile loading apparatus during μCT imaging. Thus, the contrast of a reconstructed image around the inclusion may change if there is a fatigue crack around the inclusion.

However, for every identified inclusion, no identifiable contrast change was observed before and after fatigue loading.

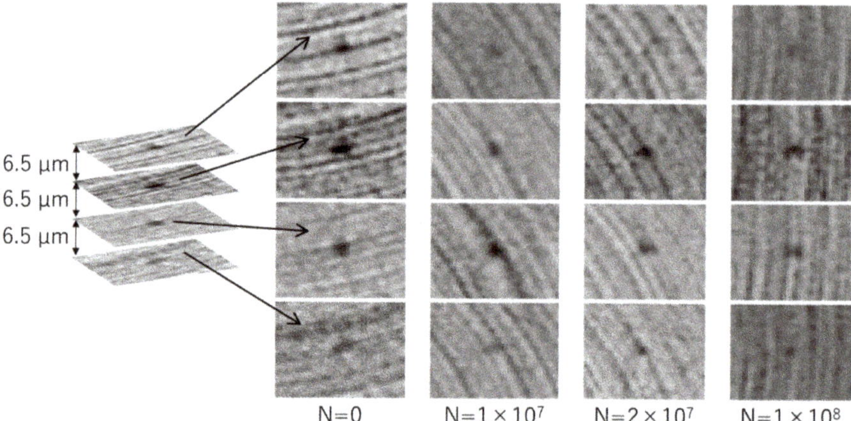

Figure 8. Reconstructed images of an inclusion #21 at every observation step.

Similarly, crack nucleation in the base material, so-called matrix cracking, was carefully inspected, but no identifiable crack nucleation was observed in the base material.

4. Discussion

As mentioned in Introduction, it is thought that most of the fatigue life is consumed by the formation of ODA and the crack growth rate in ODA is very slow, whereas penny-shaped microcracks around many inclusions nucleate in the early stage of very high cycle fatigue [4–6]. Although the internal fatigue origin was not included in the observation volume, no identifiable internal fatigue crack around every inclusion was observed and there was at least up to 70.9% of fatigue life in the imaging volume.

In this study, we could not inspect the formation of ODA from the reconstructed images after fatigue loading. There might be several possible reasons to interpret the results. The first possible reason is that there is fatigue damage around inclusions, but the size is so small that the fatigue damage cannot be detected by this set-up. The second possible reason is that fatigue damage occurs just before the final break, and so there is no damage around inclusions in the μCT images. The resolution of the reconstructed images was limited to 6.5 μm, and the incident angle of X-ray was needed to be set almost parallel to the crack plane of ODA. For more detailed in-situ-inspection of ODA formation, improvements of resolution and experimental set-up should be required.

The fatigue crack nucleated from an inclusion in this study. Thus, the fatigue strength is compared with the prediction of Murakami's approach [10], which is widely used for predicting fatigue strength for alloy steels. The diameter of the inclusion of the fatigue origin was about 20 μm, and the Vickers hardness was 435 HV. Substituting the values into the following equation for internal inclusion gives an estimate of the fatigue strength for the specimen.

$$\sigma_w = \frac{1.56(HV + 120)}{\left(\sqrt{area}\right)^{1/6}} \tag{1}$$

The predicted fatigue strength is 540 MPa, which is 110 MPa lower than the applied stress amplitude in this experiment. It is known that a higher strain rate by ultrasonic fatigue testing may results in higher fatigue strength [13]. The higher fatigue strength in this experiment was possibly caused by a higher strain rate.

5. Conclusions

In this study, synchrotron radiation μCT imaging of Cr-Mo steel subjected to cyclic loading in excess of 10^8 cycles was conducted in order to examine the feasibility of applying synchrotron radiation μCT imaging to alloy steels for the non-destructive inspection of inclusions for potential origins of internal fatigue damage in the very high cycle region.

1. The proposed experimental set-up allows us to observe inclusions of more than 10 μm nondestructively, one of which is the potential origin of an internal fatigue crack.
2. No identifiable internal damage was observed within the imaging volume at least up to 70% of the fatigue life. For more detailed in-situ-inspection of ODA formation, improvements of resolution and experimental set-up should be required.

Author Contributions: Conceptualization, Y.S., T.F. and K.A.; methodology, Y.S., K.A., T.S. and A.S.; formal analysis, S.M.; investigation, Y.S., S.M., T.F. and K.A.; writing—original draft preparation, Y.S. and S.M.; writing—review and editing, T.F. and K.T.; visualization, S.M.; supervision, Y.S. and K.T.; project administration, K.A.; funding acquisition, K.A. and T.S.

Funding: The synchrotron radiation experiments were performed at the BL22XU of SPring-8 with the approval of the Japan Synchrotron Radiation Research Institute (JASRI) (Proposal No. 2014B3723, 2015A3723, and 2016B3722).

Acknowledgments: We thank Masato Nishikawa for his great effort in preliminary experiments in SPring-8, and Yuki Shimizu, Masato Fujiwara, Ryota Nagayama, and Suzuyo Takahata for their help in the experiment in SPring-8, all of whom were students at Shizuoka University. We also thank Yoshihito Kuroshima at Kyushu Institute of Technology for his valuable suggestions and discussion.

References

1. Sakai, T. Review and prospects for current studies on very high cycle fatigue of metallic materials for machine structural use. *J. Solid Mech. Mater. Eng.* **2009**, *3*, 425–439. [CrossRef]
2. Murakami, Y.; Nomoto, T.; Ueda, T.; Murakami, Y. On the Mechanism of Fatigue Failure in the Superlong Life Regime (N > 10^7 cycles). Part I: Influence of Hydrogen Trapped by Inclusions. *Fatigue Fract. Eng. Mater. Struct.* **2000**, *23*, 893–902. [CrossRef]
3. Sakai, T.; Sato, Y.; Oguma, N. Characteristic S–N property of high carbon chromium bearing steel under axial loading in long life fatigue. *Trans. Jpn. Soc. Mech. Eng. A* **2001**, *67*, 1980–1987. [CrossRef]
4. Kuroshima, Y.; Ikeda, T.; Harada, M.; Harada, S. Subsurface crack growth behavior on high cycle fatigue of high strength steel. *Trans. Jpn. Soc. Mech. Eng. A* **1998**, *64*, 2536–2541. [CrossRef]
5. Luo, L.; Shiozawa, K. Effect of two-step load variation on super-long life fatigue and internal crack growth behavior of high carbon-chromium bearing steel. *Trans. Jpn. Soc. Mech. Eng. A* **2002**, *68*, 1666–1673. [CrossRef]
6. Ogawa, T.; Stanzal-Tschegg, S.E.; Shönbaue, B.M. A fracture mechanics approach to interior fatigue crack growth in the very high cycle regime. *Eng. Fract. Mech.* **2014**, *115*, 241–254. [CrossRef]
7. Yoshinaka, F.; Nakamura, T.; Nakayama, S.; Shiozawa, D.; Nakai, Y.; Uesugi, K. Non-destructive observation of internal fatigue crack growth in Ti-6Al-4V by using synchrotron radiation μCT imaging. *Int. J. Fatigue* **2016**, *93*, 397–405. [CrossRef]
8. Kasahara, R.; Nishikawa, M.; Shimamura, Y.; Tohgo, K.; Fujii, T. Evaluation of very high cycle fatigue properties of β-titanium alloy by using an ultrasonic fatigue testing machine. *Key Eng. Mater.* **2016**, *725*, 366–371. [CrossRef]
9. Nakai, Y.; Shiozawa, D.; Morinaga, Y.; Kurimura, T.; Okada, H.; Miyashita, T. Observation of inclusions and defects in steels by micro computed-tomography using ultrabright synchrotron radiation. In Proceedings of the Fourth International Conference on Very High Cycle Fatigue, Ann Arbor, MI, USA, 19–22 August 2007; Allison, J.E., Jones, J.W., Larsen, J.M., Ritche, R.O., Eds.; TMS: Warrendale, PA, USA, 2007; pp. 67–72.
10. Murakami, R. *Metal Fatigue: Effects of Small Defects and Nonmetallic Inclusion*; Yukendo Ltd.: Tokyo, Japan, 1993; p. 90, ISBN 978-4842593029.

11. *WES 1112: Standard Method for the Ultrasonic Fatigue Test in Metallic Materials*; The Japan Welding Engineering Society: Tokyo, Japan, 2017; p. 7.
12. Computed Tomography in SPring-8 (SP-µCT). Available online: http://www-bl20.spring8.or.jp/xct/index-e.html (accessed on 15 March 2019).
13. Mayer, H. Recent Developments in Ultrasonic Fatigue. *Fatigue Fract. Eng. Mater. Struct.* **2016**, *39*, 3–29. [CrossRef]

MDPI
St. Alban-Anlage 66
4052 Basel
Switzerland
Tel. +41 61 683 77 34
Fax +41 61 302 89 18
www.mdpi.com

Metals Editorial Office
E-mail: metals@mdpi.com
www.mdpi.com/journal/metals